TIME-MARCHING

Cranfield Series on Turbomachinery Technology

Series Editor: Robin L. Elder

Time-Marching

A step-by-step guide to a flow solver

MICHAEL LOBO
School of Mechanical Engineering
Cranfield University
Bedford

Ashgate

Aldershot • Brookfield USA • Singapore • Sydney

Published by
Ashgate Publishing Ltd
Gower House
Croft Road
Aldershot
Hants GU11 3HR
England

Ashgate Publishing Company
Old Post Road
Brookfield
Vermont 05036
USA

British Library Cataloguing in Publication Data

Lobo, Michael
    Time marching : a step-by-step guide to a flow solver. -
    (Cranfield monographs on turbomachinery and C.F.D.)
    1. Fluid dynamics - Data processing
    I. Title
    620.1'064'0285

Library of Congress Catalog Card Number: 96-85547

ISBN 0 291 39826 X

Printed in Great Britain by
Antony Rowe Ltd, Chippenham, Wiltshire

Contents

v

Preface

From its birth as a field of scientific research about two and a half decades ago, Computational Fluid Dynamics (CFD) has assumed such importance that it can no longer be regarded as just another 'branch' of fluid mechanics or numerical analysis. Whereas twenty years ago, a specific chapter on CFD might have formed part of a book on recent developments in fluid mechanics or on modern applications of numerical analysis, there are in existence today several textbooks devoted entirely to CFD and its various branches; indeed these branches are now so diverse that it is becoming increasingly difficult to do justice to all of them in a single volume. Whereas twenty years ago, a CFD Research Group might have formed part of a larger department of, say, applied mathematics or aeronautical engineering, it is now not uncommon to find departments and even institutes devoted wholly to CFD research. And to accommodate the vast number of research papers that are being written on this subject every year, new journals have had to be created. It is perhaps no exaggeration to claim that CFD has been the single most important development in applied mathematics to have emerged as a natural outcome of the high-speed computer era.

It is hardly necessary to add that academic courses in CFD, leading to post-graduate degrees are now being offered at various universities. A typical such CFD course might begin with the general mathematical background. The basic equations which govern the flow - mass, momentum and energy conservation - may be derived under different physical assumptions such as inviscidity, incompressibility, irrotationality, etc. Moreover they may be expressed in a framework of Cartesian, polar or general curvilinear coordinates. This may be followed by a more general discussion of partial differential equations, the fundamental differences between parabolic, hyperbolic and elliptic PDEs, (from both an analytic and numerical point of view) and the physical circumstances under which the governing equations may fall into one or other of these categories.

Having dealt with the fundamentals, the CFD course might next move on to elementary solution techniques for model problems. A typical solution technique involves two distinct steps: first discretization of the governing PDEs into a system of linear or (more often) nonlinear algebraic equations, and second the numerical solution of this algebraic system. The first step would have to make use of discretization techniques such as finite difference, finite element or finite volume. The second step may be handled in any one of a number of different ways depending on the complexity of the algebraic system and the mathematical nature of the original governing equations. Here the student will learn about direct and iterative approaches, explicit and implicit approaches, the positive and negative features of these different approaches, various circumstances under which they may or may not be applicable, and a few other important topics integrally associated with any numerical method of solution such as stability, consistency and convergence.

The topics covered thus far ought to suffice for a basic course in CFD. An advanced course must naturally probe a little deeper: model problems should be replaced by more realistic problems and emphasis must be placed on solution techniques that are assuming increasing importance in current research. Now the most noticeable development over the last two decades has been the tremendous increase in power of the digital computer as a result of which it has now become possible to deal numerically with the complete system of Navier-Stokes equations. Nevertheless solution techniques for lower-order systems such as the Euler and even the potential equations should not be regarded as obsolete - in fact, as recently as ten to fifteen years ago, they were actually in the forefront of CFD research.

An advanced course, therefore, might begin by summarising the solution techniques which were popular in the 1970s for dealing with the subsonic and (especially) the transonic small perturbation and full potential equations. It might then move on to the early 1980s - the 'heyday' of the Euler solvers. This is, incidentally, an appropriate moment to point out that some of the most important methods in use today for handling the complete set of Navier-Stokes equations originally made their appearance in the context of the Euler equations. These include the so-called time-marching technique - the subject of the present work.

And so on to the late 1980s and the early 1990s - the 'modern era' marked by the domination of Navier-Stokes solvers. Even the most advanced of CFD courses cannot really hope to cover this area in much depth; it properly belongs to the realm of research. Characteristic problems associated with Navier-Stokes solvers - such as turbulence modelling - must of course be discussed, and general classes of solution techniques such as 'time-marching' and 'pressure-correction' can be studied in just a little detail. Aids to convergence, acceleration techniques, and other refinements can just about be touched upon.

This brings us to an all-important question. Will a student who has attended a basic or even an advanced course in CFD be in a position to commence research in the area - research that is of significance in the context of present-day industrial requirements or even simply in the context of present-day academic interest?

In order to answer this question fairly, one must first reflect that research in CFD generally amounts to the running of a CFD code and the analysis of its results. But - and here we come to the crux of the question - whereas fifteen years ago, CFD codes (generally potential solvers) were developed from 'scratch', this is rarely true today. The codes which are at the forefront of present-day research - those capable of handling the unsteady compressible three-dimensional Reynolds-averaged Navier-Stokes equations - are the culmination of several years of development. Most of them were probably 'born' in the early 1980s, first as two-dimensional Euler solvers, and then 'grew' monotonically over the years so as to be able to deal with successively higher orders of governing equations. And naturally refinements, such as multigridding for the purpose of accelerating convergence, would have found their way into the code somewhere along the line.

To return to our all-important question. A student who has attended an advanced course in CFD should be in a position to commence research in this area *provided he has access to an already developed CFD code which he may then use as a basis for further refinement.* If he is not in this fortunate position, it would be extremely difficult for him to compete with those workers who have at their disposal codes which have been developed for a decade or more. After all, even an advanced course in CFD can do no more than lay out the principles of time-marching, pressure-correction and other important present-day solution techniques for handling the Navier-Stokes equations. But from the understanding of principles to the development of a stable robust CFD code is an enormous step indeed.

Considered from this viewpoint, it would seem a fairly good idea for some of the world's leading CFD codes to be made available to academic researchers in general. But although this would undoubtedly aid the overall progress of research, 'copyright holders' of CFD codes are understandably reluctant to reveal their secrets for fear of commercial misuse. As a result, enterprising young CFD researchers, especially at lesser-known universities, have no choice but to develop their codes from the beginning. And while it is no doubt true that the writing and development of this CFD code would afford the enterprising young researcher a lot of personal satisfaction, it is most unlikely that the final code (which must of course be completed within the period over which the student works on his doctorate) will be able to reach out to new frontiers of research.

In the latter half of the 1980s, two young Taiwanese students Wen Chang Tsay and Chung Pin Cheng joined the School of Mechanical Engineering at Cranfield and worked towards a Ph.D. degree in the field of CFD. At the time they joined the school, the CFD Research Group was barely out of its infancy; a pressure-correction code was in the process of development, but no work of note had been accomplished in time-marching. Partly in order to complement the group activities, Tsay and Cheng opted to work in time-marching rather than pressure-correction. They had, of course, to start from scratch. Tsay developed a time-marching code based on ideas set forth by W.Dawes at Cambridge, and Cheng developed a time-marching code based on ideas set forth by A.Jameson at Princeton. Needless to add, these ideas could at best have been of as much assistance as an advanced course in CFD.

Over the next few years, Tsay and Cheng developed their respective time-marching codes, the former working from 1986 to 1989 and the latter from 1988 to 1991. In this short space of time they were able to produce eminently satisfactory codes which, while perhaps not quite in the category of frontier-breaking research, were certainly capable of accomplishing the tasks for which they were designed. Indeed the results from some test cases indicated that the codes could, on occasion, actually perform more efficiently than their (internationally renowned) counterparts at Cambridge and across the Atlantic.

I had the good fortune to have been appointed a joint supervisor (along with Professor R.L. Elder, the head of our department) for both W.C.Tsay and C.P.Cheng. My other commitments over the years prevented me from taking a very active role in their research and it is only fair to say that the codes they developed were almost entirely their own efforts. Both Tsay and Cheng have now left Cranfield and I would like to feel that their work is not lost to posterity.

And this brings me to the subject of the present book: it is, in essence, a manual on a time-marching code. For reasons of time, it will not be possible for me to produce a book on the work of both Tsay and Cheng and I have chosen the CFD code of the latter partly because it is more recent and partly because I personally spent some time in analysing it in connection with some of my own researches on the acceleration of iterative schemes. I might add that the form in which the code is presented here is considerably different from Cheng's original; I have deliberately re-organized it in such a manner so as to make it easily understandable to students. I have also put in a few refinements of my own.

This brings me to my second reason for writing this work. For reasons explained earlier, there are at present few if any books on existing CFD codes that are accessible to the academic world in general. And yet such works are of extreme importance if one is to bridge the gap between a CFD course for post-graduate students and the frontiers of current research. In their time Tsay and Cheng had to start their research at the very beginning; I sincerely hope this will not have to be said for future students - at least in the field of time-marching.

A few general comments. Although a reader of this work would find it easier to follow if he already has some background in CFD, this is not essential. The only pre-requisite to an understanding of this work is a sound knowledge of engineering mathematics. We start from the governing equations, explain the theory behind the time-marching approach and proceed step by step until we are in a position to list the complete computer program for the Euler solver in two dimensions. The present work will be restricted to two dimensions partly for resons of time, but chiefly because in the first instance ideas can be assimilated much more easily in the context of two dimensions. A three-dimensional extension is then merely a forward step with no real change in 'strategy'. A companion volume on the full Navier-Stokes solver and on the three-D extension may appear later.

One final comment. It is customary to begin works of this nature with an extensive literature survey, but I shall skip this because it is not really relevant to an understanding of what follows. Readers interested in recent developments in the field of time-marching need merely press this key word on the computer systems in their library (yes, 'time-marching' has now achieved the status of a general library key word). This will immediately put them abreast of the available literature and a glance at a couple of these will lead to additional references - and so on. When the readers have had their fill of the literature survey, they can commence their study of the present work - or, better, vice versa!

Michael Lobo
June 1996

Figures and Abbreviations:
A brief note

The text contains fourteen figures - one in Chapter 2 and thirteen in Chapter 3. As the figure captions are of significance only in the context of the associated text, they are not listed separately here. Note that all figures appear on odd-numbered pages; specifically, each figure is placed on the first odd-numbered page following its original reference. This makes it easier for the reader to refer simultaneously to the figures and to the relevant text.

Few abbreviations are employed; in general these are standard and well-known to students of applied mathematics: for example, ODE (Ordinary Differential Equation), PDE (Partial Differential Equation), IVP (Initial Value Problem), etc. Then, of course, there is CFD (Computational Fluid Dynamics) itself. One other is CFL, an acronym for the Courants-Friedrichs-Lewy criterion in stability theory.

1 Time-Marching: General mathematical concepts

In the context of Computational Fluid Dynamics, the first effective application of time-marching is generally regarded as that by Moretti and Abbett (1966) - and in the context of turbomachinery in particular, that by Denton (1974). However, the principles underlying the time-marching approach are virtually as old as numerical analysis itself and can be applied to a very wide range of differential equations starting from the simplest first order Ordinary Differential Equations (ODEs). Consider, for example, the first order Initial Value Problem (IVP):

$$y' = f(x,y) \qquad \qquad \text{...(1.1a)}$$
$$y(x_0) = y_0. \qquad \qquad \text{...(1.1b)}$$

The classical approach to solving this problem numerically is to approximate the derivative y' by the first-order difference:

$$y'(x) = [(y(x+h) - y(x)] / h \qquad \qquad \text{... (1.2)}$$

so that

$$y(x+h) = y(x) + h\{f(x,y)\}. \qquad \qquad \text{... (1.3)}$$

Because $y(x_0)$ is known from (1.1b), we may proceed to determine $y(x_0+h)$, and by repeated applications of Equation (1.3), $y(x_0+2h)$, $y(x_0+3h)$, etc. In other words we are solving the problem by *marching* forward in steps of h . Thus the classical method of solution could be referred to as a marching[1] approach and, indeed, if the independent variable x could be identified with time, the solution procedure would come under the category of *time-marching* (the interval h being the *time-step*).

Note, incidentally, that the classical method of solution shown above is only of first order accuracy. It is of course possible to increase the order of accuracy of the

---

[1] According to some authorities, the use of the word 'marching' as a solution technique was first proposed by L. F. Richardson, a British meteorologist and numerical analyst who flourished during the early years of the present century. His pioneering work on numerical solutions of partial differential equations is sometimes regarded as heralding the dawn of CFD.

solution algorithm; indeed the first-order method of solution is used mainly by way of introducing the subject in a classroom lecture; if a solution to the IVP (1.1) is desired for a practical purpose, the numerical analysis almost invariably makes use of a high-order method such as the Runge-Kutta. A high-order algorithm generally involves a certain number of intermediate calculations before $y(x+h)$ can be expressed in terms of $y(x)$ and the function $f(x,y)$. These intermediate calculations form part of the step between x and $x+h$, but the *step* is still present. In other words the problem is still being solved by *marching* forward, except that the rate of marching is slower; more effort is being put into the computation over the same distance - or, if the independent variable is time, more effort is being put into the computation over the same *time-step*.

Let us now move on to a differential equation of the second order:

$$y'' = f(x,y,y') \qquad \qquad \ldots (1.4a)$$
$$y(x_0) = y_0 \qquad \qquad \ldots (1.4b)$$
$$y'(x_0) = y_0'. \qquad \qquad \ldots (1.4c)$$

Once again we are considering an initial value problem. It may be solved numerically by a classical (i.e. classroom) first-order first-order scheme:

$$y(x+h) = y(x) + h\{y'(x)\} + \tfrac{1}{2}h^2\{f(x,y,y')\} \qquad \ldots (1.5a)$$
$$y'(x+h) = y'(x) + h\{f(x,y,y')\}, \qquad \qquad \ldots (1.5b)$$

or by a higher order scheme such as the fourth-order Runge-Kutta-Nystrom method. But whatever be the accuracy of the scheme employed, we are still solving the problem step by step, starting from the initial point (or instant) x_0 and moving to x_0+h, x_0+2h, etc. The main additional complication (in comparison to the numerical solution of first order equations) is that both y and y' must now be computed at each step[2]. But the principle underlying the solution procedure is the same for both the IVPs (1.1) and (1.5) - the *marching* forward in steps of h. And if the independent variable is time (and if the reader will pardon this repetition), both IVPs (1.1) and (1.5) are being solved by *time-marching*.

Thus far we have only been considering initial value problems. But what if the second order differential equation is described by boundary rather than initial conditions:

$$y'' = f(x,y,y') \qquad \qquad \ldots (1.6a)$$
$$y(x_0) = y_0 \qquad \qquad \ldots (1.6b)$$
$$y(x_1) = y_1 \qquad \qquad \ldots (1.6c)$$

---

[2] More precisely, in the solution procedure for (1.1), y is being solved for by numerical approximation whereas y' is obtained by straightforward evaluation of $f(x,y)$. In the solution procedure for (1.5), both y & y' are being solved for by numerical approximation whereas y'' is determined from $f(x,y,y')$.

and the solution is desired in the domain between x_0 and x_1.

In the first place we should recall that for arbitrary values of y_0 and y_1, there is no guarantee of the existence of a solution to the problem - and should a solution exist, there is no guarantee that it will be unique. However questions of existence and uniqueness do not concern us here. Assuming the problem is indeed well-posed, how do we go about finding a solution? Unlike the initial value problem there is no standard procedure. One method of attack (probably the best) is to discretize the interval between x_0 & x_1, then approximate the differential equation (1.6) by a difference equation at every discrete point, and finally solve the resulting algebraic system of equations by an iterative procedure. In a solution procedure of this nature, there is no question of marching involved. In an iterative procedure, one returns to the starting point at the end of each iteration (until one is satisfied that the iterations have converged). *But in a marching procedure there is never any return. One marches forwards only.*

It would seem then that marching procedures are characteristic of initial value problems. And - to come to a significant point - if the independent variable is time, the problem is almost invariably an initial value problem because the dependent variable under consideration can in general only be specified at some past instant and not in the future[3]. *Thus general marching methods may not always be appropriate for general differential equations, but time-marching methods are almost invariably suitable for time-dependent differential equations.*

This brings us to an interesting observation. Every student of applied mathematics should be aware that boundary value problems are, in general, much more difficult to handle than initial value problems. Furthermore, as we have just seen, initial value problems can be handled by marching techniques whereas boundary value problems require iterative techniques[4]. Can we then draw the inference that *marching* is simpler than *iteration*, or, to put it another way, that iterative approaches are more advanced?

Let us frame this question in a slightly different manner. Let us designate one solution technique, say A, as *superior* to another, say B, if A can be used for all types of problems to which B is applicable as well as to a wider class of problems. Now the marching which characterizes an initial value problem can be considered a special case of an iterative scheme - one in which convergence has taken place at the very first iteration. It would appear therefore that iterative techniques are superior to marching techniques in the sense described above because they may be applied to both initial and boundary value problems, whereas marching may

---

[3] On the other hand one might specify a variable at a given instant and investigate its past behaviour (for example that of a planetary configuration). This involves marching backward in time. However this does not contradict what we have been saying; the past has simply become the direction forward.

[4] There are other means of handling boundary value problems, but they do not really make the solution procedure any simpler.

properly be applied only to the former.

But the very simplicity of the marching approach provides it with advantages which may not be obvious at first sight. Consider, once again, the boundary value problem (1.6). Let us complexify the problem a little, by treating (1.6) as a special case of a Partial Differential Equation (PDE):

$$y_t = y_{xx} - f(x,y,y_x) \qquad \qquad \dots (1.7a)$$
$$y(x_0,t) = y_0 \text{ for } t \geq 0 \qquad \qquad \dots (1.7b)$$
$$y(x_1,t) = y_1 \text{ for } t \geq 0 \qquad \qquad \dots (1.7c)$$
$$y(x,0) = g(x). \qquad \qquad \dots (1.7d)$$

Obviously if the dependence on t is ignored, Eqs (1.7) reduce to Eqs (1.6). For the moment, let us not concern ourselves with the function g(x) other than to point out that it must of course satisfy the conditions $g(x_0) = y_0$ and $g(x_1) = y_1$.

Students of applied mathematics immediately ought to recognize (1.7) as a classical example of a parabolic partial differential equation. It is indeed so fundamental that it forms the starting point of many text books on numerical solutions to PDEs both at an elementary level (for example, G.D. Smith (1965)) as well as at a more advanced level (for example, Richtmyer and Morton (1967)). Basically the procedure is as follows. The domain between x_0 and x_1 is suitably discretized and for a fixed value of t, the differential terms in x are approximated by appropriate difference formulae. This essentially amounts to a system of *ordinary differential equations* in t which must be solved at every discrete point. Furthermore, each of these ODEs are *initial value problems* and may therefore be solved by a marching procedure. Of course each ODE in the system is dependent on some of the others and so cannot be solved independently. But the entire system can be solved simultaneously for all values of x at successively increasing levels of t. We are marching in t; we are t-marching!

For the particular case when the function f is identically zero, Equations (1.7) represent the variation of heat conduction along a rod stretching from x_0 to x_1 and insulated at either end. In this case the variable t literally denotes time and the answer to the problem will provide the value of the temperature at all points along the rod at all instances of time after the initial instant when the heat distribution is known. This answer is of course obtained by time-marching from the initial instant. Of course the marching procedure will have to come to a halt at some point of time and in practice this may be done when the temperature distribution over the rod is found to have attained a *steady state*, that is, a point at which further time steps have little if any effect on the temperature distribution. From this moment onwards, the time dependent terms can effectively be set to zero, so that in the case of the heat conduction problem under consideration, the temperature distribution will now satisfy the trivial differential equation,

$$y_{xx} = 0 \quad ; \quad y(x_0) = y_0 \quad ; \quad y(x_1) = y_1 \qquad \qquad \dots (1.8)$$

4

whose solution is:

$$y = y_0 + \{(x-x_0)/(x_1-x_0)\} (y_1-y_0).\qquad \ldots (1.9)$$

Note that the time marching procedure will eventually have to arrive at this steady-state solution regardless of the initial temperature distribution[5].

For the general case when the function f is not identically equal to zero, the steady state solution of Equations (1.7) will satisfy Equations (1.6). We have in fact pointed out an alternative method for solving the Boundary Value Problem (1.6) - a method which does make use of the marching (shall we say time-marching?) principle. And this brings us back to the question we posed earlier - about the relative superiority of the two solution techniques, iteration and marching. Our first look appeared to suggest that iterative methods were applicable to a wider class of problems, but our second look has left the question open again.

Admittedly the reader might raise an objection to the solution procedure we have just described; the idea of converting an ordinary differential equation into a partial differential equation just so as to be able to implement a marching algorithm appears a little far-fetched. In principle, this criticism is justified; our idea was merely to show that marching schemes *could* be used, even for problems where they are traditionally considered inapplicable. Having said this, it should also be added that the idea of complexifying a system in order to be able to implement a marching procedure is not as far-fetched as it might appear. For a particular set of equations, a marching procedure, if applicable, is in general a whole order of magnitude more efficient than an iterative procedure because it is tantamount to an iterative procedure which terminates at the very first iteration. Thus even if one were to add a whole new dimension to a system of equations in order to be able to implement a marching procedure, the algebraic operations involved need not be greater than those required by the original system and can conceivably be less. Indeed, it is worth pointing out that in some special instances the two approaches can follow identical paths. Consider Equation (1.6a). A central difference approximation yields

$$[y(x+\Delta x)-2y(x)+y(x-\Delta x)] / \Delta x^2 = f(x,y,y')\qquad \ldots (1.10)$$

which can be written iteratively as

$$y_{n+1}(i) = \tfrac{1}{2}[y_n(i+1) + y_n(i-1) - \Delta x^2\{f(i)\}].\qquad \ldots (1.11)$$

Here the suffix n denotes the iteration number and the argument i denotes the index number of the discrete point which is currently being operated upon in the iterative cycle. For additional generality, one may replace (1.11) by

---

[5] A purist might find fault with this statement; for example numerical instabilities may be present in the time-marching scheme and it may fail to arrive at any solution. These matters will be discussed in due course, but let us first understand the underlying principles.

$$y_{n+1}(i) = \alpha[\frac{1}{2}\{y_n(i-1)+y_n(i+1)-\Delta x^2 f(i)\}] + (1-\alpha)y_n(i) \quad \ldots (1.12)$$

Here α is a *relaxation factor* and may be chosen so as to accelerate convergence. After choosing some initial guess for y_0, the iterative procedure can commence.

Now consider Equation (1.7a). Backward differencing in t and central differencing in x yields:

$$[y(x,t+\Delta t)-y(x,t)]/\Delta t = [y(x+\Delta x,t)-2y(x,t)+y(x-\Delta x,t)]/\Delta x^2 - f(x,y,y_x)$$
$$\ldots (1.13)$$

If we now use the suffix n to denote the nth time level (and the argument i as before), Equation (1.13) may be written as

$$y_{n+1}(i) = (1-2r)y_n(i) + r[y_n(i+1)+y_n(i-1)-\Delta x^2 f(i)]. \quad \ldots (1.14)$$

where $r = \Delta t/\Delta x^2$. Clearly the schemes (1.12) and (1.14) are identical if $r = \frac{1}{2}\alpha$. If the function g(x) in Equation (1.7d) can be identified with the initial guess for the iterative approach, then the numerical steps en route to the solution will be precisely the same regardless of whether one employs the iterative procedure (1.12) or the time-marching scheme (1.14)[6].

It is now time to review what we have learnt so far about the concept of time-marching. Our first look suggested that it was applicable to only a specific class of problems and did not have the generality possessed by iterative procedures. Then we took a second look. This revealed that time-marching had hidden potentialities and could be applied to a wider class of problems by introducing an artificial time-dimension. This led to a comparison with the iterative approach and a detailed analysis (in fact a third look) appeared to show that the artificial time-marching procedure was nothing more than an alternative means of deriving the iterative system of equations (rather than a distinct solution procedure).

What then can we conclude about time-marching as a solution procedure? If we claim that we are solving the steady Euler or Navier-Stokes equations by an artificial time-marching algorithm, are we really fooling ourselves, when all we are doing is providing a high-sounding name to an old established iterative procedure? Not so! In fact it is now time to take a fourth look at time-marching.

Let us open the curtain on Scene 4 by viewing two very fundamental problems in the theory of hyperbolic partial differential equations - the so-called 'linear advection equation' and its nonlinear counterpart which we shall refer to as the 'nonlinear advection equation'. In both problems, t may be identified with time.

---

[6] This assumes of course that the same operative procedure is used for both schemes. For example, a point-by-point solution (Jacobi iteration) of (1.12) would be equivalent to an *explicit* solution of (1.14). A simultaneous solution of (1.12) (using the tri-diagonal matrix algorithm) would be equivalent to an *implicit* solution of (1.14).

Linear advection equation:

$$u_t + cu_x = 0 \qquad \dots (1.15a)$$
$$u(x,0) = u_0(x) \qquad \dots (1.15b)$$

Nonlinear advection equation:

$$u_t + uu_x = 0 \qquad \dots (1.16a)$$
$$u(x,0) = u_0(x) \qquad \dots (1.16b)$$

A few words of explanation: basic courses (and text books) on partial differential equations invariably deal with the classification of second-order PDEs - thus the PDE

$$au_{xx} + bu_{xy} + cu_{yy} + du_x + eu_y + f = 0 \qquad \dots (1.17)$$

is hyperbolic, parabolic or elliptic according to whether the sign of the discriminant b^2-4ac is greater than, equal to, or less than zero. For a student, this rule is very easy to remember as it is precisely analogous to the classification of second-order algebraic equations into hyperbolas, parabolas or ellipses in elementary analytic geometry. But what of first-order PDEs? These have no counterpart in analytic geometry. Indeed a student may at first be inclined to regard equations such as (1.15a) and (1.16a) as special cases of second-order PDEs with $a = b = c = 0$, and accordingly classify them both as parabolic.

The point to be remembered here is that PDEs (and indeed anything else) should be classified not in accordance with the blind application of a formula (whose applications may be restricted) but in accordance with the characteristic features of the problem. Thus the solution to a hyperbolic PDE is characterized by curves (appropriately called characteristics) along which the problem reduces to an ordinary differential equation. It is also characterized by a dependence on only a part of the initial data. By contrast the solution to a parabolic PDE depends on the entire initial data and part of the boundary data (if any). The solution to an elliptic problem depends only on boundary data - which must be specified as a closed curve. There are other such fundamental distinctions between the different classes of PDEs.

We can now see why Equations (1.15a) and (1.16a) are hyperbolic - they possess the characteristic features of a hyperbolic problem, i.e. dependence on only a part of the initial data and the existence of characteristic curves. Indeed in the case of Equation (1.15a), the characteristic curves are the straight lines (in the x-t plane) 'x = ct' along which the solution is constant. Any point on this line will depend only on the initial data at its point of intersection with the x-axis (and in fact will be precisely equal to it). In the case of Equation (1.16a), the characteristic curves are also straight lines, but their slopes will vary in accordance with the initial data.

General rules for classification of first-order and systems of first-order equations are likely to be found in more recent text books on PDEs because the subject is gaining in importance. An interested reader may refer, for example, to

Zachmanoglou and Thoe (1976), Chapter 10. The treatment is at a fairly elementary level. For a deeper study, one may refer to Garabedian (1964), Section 3.5.

For those readers who are already well-acquainted with these general rules of classification, I offer my apologies for this digression. Let us now proceed to examine the IVPs (1.15) and (1.16) in the context of time-marching.

Consider first (1.15). This is the simplest possible hyperbolic IVP; in fact its analytic solution

$$u = f(x-ct) \qquad \qquad \ldots (1.18)$$

appears so trivial that a student may be pardoned for supposing that the problem is not worth worrying about at all! However the analytic solution is only one side of the picture. The problem can also be tackled numerically - not for the purpose of obtaining a numerical solution, but for the purpose of testing different numerical techniques and comparing them with the known analytic solution. And there is virtually no limit to the number of different numerical techniques that can be tested on this simple equation. There are numerical schemes involving two time levels, schemes involving three or more time levels, explicit schemes, implicit schemes, schemes of first-order accuracy in x, of second-order accuracy, of third-order accuracy and so on. Moreover the schemes may involve forward, central or backward differencing or a judicious combination. There are several other parameters which may be varied - too numerous to mention here. An interested reader may refer, for example, to the text book *Numerical Solution of Partial Differential Equations* by Lapidus and Pinder (1982), Section 6.2, or to the text book *Computational Fluid Mechanics and Heat Transfer* by Anderson, Tannehill and Pletcher (1984), Section 4.1, or, for a more theoretical discussion, to the report by Roe (1981). Here we shall just make one relevant comment: All these widely varying schemes have a common feature. They use the principle of time-marching.

The importance of the study of numerical solution techniques for the linear advection equation simply cannot be overestimated because it provides an opportunity to analyse, at a basic level, many of the difficulties encountered in the numerical treatment of more realistic hyperbolic problems. These include global features such as stability of the numerical scheme as a whole as well as local difficulties which may arise, for example, due to the presence of a shock wave. For the case of the linear advection equation, a shock wave may be modelled by incorporating a discontinuity in the initial data. This discontinuity will be carried along the characteristic curve x = ct, but as the numerical schemes do not actually solve the problem along the characteristic curves[7], a certain amount of inaccuracy

---

[7] Admittedly, it is possible to formulate a numerical scheme for solving along the characteristic curves, but for general hyperbolic problems, they can become extremely cumbersome to program - even assuming the characteristic curves have been properly computed, in itself a nontrivial task. Most practical hyperbolic solvers therefore approach the problem in the 'normal' manner, i.e. setting up a grid and approximating the PDE at each grid point.

is unavoidable. If this inaccuracy were merely quantitative, it might be acceptable; however it is also qualitative. More explicitly, 'orthodox' numerical algorithms of second or higher order exhibit spurious oscillations in and around the shock - there is actually a theorem too this effect attributed to Godunov (1959). A great deal of research has gone into the understanding of how these spurious oscillations[8] arise and how they may be minimised if not eliminated. And of course this research must be carried out, at least to start with, on the linear advection equation.

Further details, and references to relevant work in the area, may be found in the review article by Roe (1982). And for those readers who may feel that all this is a considerable digression from the subject matter of the present work, it must be emphasized that this is not the case because the theory behind the time-marching code which forms the subject of the present work eventually rests on the secure foundations built up by research into the linear advection equation.

If Equation (1.15) is the most basic of all hyperbolic problems, Equation (1.16) is certainly the most basic of all *nonlinear* hyperbolic problems and may be used for a fundamental study of nonlinear effects which naturally cannot be simulated by a numerical approximation to Equation (1.15). Perhaps the most important of these is the generation of shocks from *continuous* initial data. Thus if shocks are to be modelled using a numerical algorithm to Equation (1.15), it becomes necessary to begin with a discontinuous solution i.e. as part of the initial data. However the exact solution of Equation (1.16) may possess discontinuities even when the initial data is continuous everywhere - this arises when two characteristics intersect. The best method of capturing these shocks numerically is still a subject of intensive research - and it is especially relevant in the context of the present work because any attempt to solve the Euler or Navier-Stokes equations using an artificial time-marching algorithm ought to work with a simple (and of course continuous) initial guess, if it is to be of practical use.

Incidentally Equation (1.16) is often referred to as the 'inviscid Burgers' equation' after Burgers (1948) who analysed the more general equation:

$$u_t + uu_x = vu_{xx}. \qquad \ldots (1.19)$$

Equation (1.19) (which is the simplest to combine nonlinearity with viscosity) is appropriately known as Burgers' equation, but in the opinion of the present author, the term 'inviscid Burgers' equation' for Equation (1.16) is perhaps a little unfortunate, as it would appear to regard it as just a special case of Equation (1.19) for which the viscosity v is zero, whereas in reality Equation (1.16) is so fundamental that it certainly deserves an appellation in its own right. Hence it is referred to here as the 'nonlinear advection equation', but it is just as well to point

---

[8] These spurious oscillations are known as dissipative errors and generally appear whenever the dominant term of the truncation error contains an even derivative. By contrast, if the dominant term contains an odd derivative, the solution tends to exhibit what are called dispersive errors, i.e. damping of a sharp shock. See, for example, Anderson, Tannehill & Pletcher (1984), page 92.

out that this terminology is not standard[9].

Having said this, it should also be added that from a purely numerical point of view, there is some justification in looking upon Equation (1.16) as a special case of Equation (1.19) because extensive research has shown that it is very difficult to obtain numerical solutions to Equation (1.16) in the presence of shocks unless a small diffusion term is added to the right hand side of the equation[10]. This small diffusion term - appropriately called 'artificial viscosity' -was first proposed in a classic paper by Von Neumann and Richtmyer (1950). In effect, the numerical algorithm is now dealing with Equation (1.19), though it is seeking a solution to Equation (1.16).

Methods for treating the linear advection equation (1.15) and the nonlinear advection equation (1.16) may be extended first to general first-order hyperbolic equations of the form:

$$u_t + \{f(u)\}_x = 0 \qquad \qquad \text{... (1.20a)}$$

or equivalently (with $g = f_u$)

$$u_t + g(u)u_x = 0 \qquad \qquad \text{... (1.20b)}$$

and then to systems of first-order hyperbolic equations:

$$\mathbf{U}_t + \mathbf{F}_x = 0 \qquad \qquad \text{... (1.21a)}$$

or equivalently

$$\mathbf{U}_t + [\mathbf{F'}]\mathbf{U}_x = 0 \qquad \qquad \text{... (1.21b)}$$

where $\mathbf{U}$ and $\mathbf{F}$ are vectors having (say) n components and $[\mathbf{F'}]$ is an nxn matrix, whose components f'_{ij} are given by

$$f_{ij} = \partial f_i / \partial u_j. \qquad \qquad \text{... (1.22)}$$

The system of equations (1.21b) is hyperbolic at a point (x,t) if the eigenvalues of the matrix [G] (known as the Jacobian matrix) are all real and distinct at that point. These eigenvalues will in fact define the characteristic directions and the system of PDEs reduces to a system of ODEs along these characteristics. There does not appear to be any unanimous agreement about classification of systems where the eigenvalues are not all real and distinct, though if all are complex the system is

---

[9]   After writing this chapter, the present author had occasion to refer to the excellent text book *Linear and Nonlinear Waves* by G.B.Whitham (1974) and was pleasantly surprised to find his approach very similar.

[10]   However in some cases, the diffusion term may be implicitly present in the numerical scheme.

certainly elliptic because the lack of any real characteristic directions implies a dependence upon all the information at the boundaries. If some eigenvalues are real and some are complex, the system will possess some hyperbolic features and some elliptic features. For more details (but still at an elementary level) see the books by Zachmanoglou and Thoe (1976), Chapter 10, or by Anderson, Tannehill and Pletcher (1984), Section 2.5.

If expressed in the form (1.21a), the system is said to be in conservation form, because if **U** denotes a set of quantities (such as mass, momentum and energy) which must be conserved according to some physical law, then it ought to be possible to represent that system of physical laws in the form (1.21). Appropriately enough, such systems are designated 'Hyperbolic Conservation Laws' and have been subjected to exhaustive study for the past four decades - starting from the classic paper by Lax (1954) where he introduced the concept of 'weak solutions' (which admit discontinuities) and discussed in depth the conditions required for the existence and uniqueness of such solutions. Much of the work of Lax (and associates such as Richtmyer) is highly theoretical[11] and, admittedly, a thorough understanding of it is not essential for the engineering student who merely wishes to be able to solve a particular set of conservation laws - such as the Euler equations. Nevertheless, the fact that it has been found possible to solve the Euler equations *as a system of conservation laws* is largely due to the pioneering efforts of these theoretical researchers.

Several problems of practical interest can be expressed as a system of hyperbolic conservation laws of the form (1.21), for example the one-dimensional equations of gasdynamics, magnetogasdynamics, electromagnetic waves, river waves (i.e shallow water theory), seismic waves, traffic flow, etc. Interested readers may refer to the text books by Duffy (1986), Farlow (1982), Gustafson (1980) and, especially, Whitham (1974). There is also a recent monograph: Leveque (1990). Here, as a stepping stone to the main subject matter of this work, we shall merely present the unsteady isentropic one-dimensional Euler equations[12].

$$\rho_t + u\rho_x + \rho u_x = 0 \qquad \dots (1.23a)$$
$$u_t + uu_x + (1/\rho)p_x = 0. \qquad \dots (1.23b)$$

Equation (1.23a) expresses conservation of mass and Equation (1.23b) conservation of momentum. Three dependent variables are involved, the density ρ, the velocity u and the pressure p, but under the assumption of isentropicity,

---

[11] Peter Lax later became director of the Courant Institute of Mathematical Sciences, New York; he and his colleagues such as Paul Garabedian (who wrote a classic text book on partial differential equations) have done a great deal to put the subject on a sound mathematical footing.

[12] Of course, the one dimensional Euler equations are of significance only in the context of unsteady flow.

$$p/\rho^\gamma = A \text{ (a constant)}, \qquad \dots (1.24)$$

where γ is the ratio of specific heats, the pressure p can be eliminated and by introducing the speed of sound c, defined by

$$c^2 = dp/d\rho = \gamma A \rho^{\gamma-1}, \qquad \dots (1.25)$$

the system (1.23) can be written as

$$\rho_t + u\rho_x + \rho u_x \quad = 0 \qquad \dots (1.26a)$$
$$u_t + uu_x + (c^2/\rho)\rho_x \quad = 0. \qquad \dots (1.26b)$$

Clearly the system (1.26) can be put in the form (1.21) with

$$U = \begin{bmatrix} \rho \\ u \end{bmatrix} \quad : \quad F = \begin{bmatrix} \rho u \\ \tfrac{1}{2}u^2 + c^2/(\gamma-1) \end{bmatrix} \qquad \dots (1.27a)$$

$$[G] = \begin{bmatrix} \rho & u \\ u & \dfrac{c^2}{\rho} \end{bmatrix}. \qquad \dots (1.27b)$$

and
The eigenvalues of the matrix **[G]** are

$$\lambda_+ = u + c \qquad \dots (1.28a)$$
$$\lambda_- = u - c \qquad \dots (1.28b)$$

and the system is therefore hyperbolic.

Various numerical techniques may be used to solve the system (1.26). Zachmanogloe and Thoe (1976), Section 10.5, describe a method of solving along the characteristics - whose slopes are defined by (1.28) - using a transformation which yields two quantities r & s (Riemann invariants), each of which is constant along a particular characteristic. Because the code which forms the subject matter of the present work does not (directly) make use of the method of characteristics, we shall not go into further details, but will simply make the general remark that whether the system (1.26) is solved using characteristics or by a straightforward time-marching numerical scheme, the structure of the solution algorithm rests on fundamental research carried out for hyperbolic conservation laws.

This broad observation remains perfectly valid at all steps of the hierarchy - that is, regardless of whether we are dealing with the Euler or Navier-Stokes equations and regardless of whether we are operating in one, two or three dimensions. The only point to be remembered is that the governing equations should be expressed in their *unsteady* formulation, because only then can we be certain of the all-important requirement of hyperbolicity[13]. And this is an appropriate moment to recall the primary objective of the present work - the development of a program to solve the *steady* Euler and Navier-Stokes equations. If the time-marching technique is to be adopted (as is our aim) then an artificial time dimension will have to be introduced and the resulting unsteady flow equations must be solved until an asymptotically steady state is reached. The procedure is basically similar to the use of a time-marching technique for the initial value problem (1.7) in order to arrive at a solution for the boundary value problem (1.6).

And this naturally brings us back to 'Scene 3' where we compared an artificial time-marching procedure, for a model problem, with a straightforward iterative approach and found they were the same. What more have we learnt from 'Scene 4'?

In the first place we have learnt that any time-marching procedure, whether 'natural' or 'artificial' rests on a solid body of fundamental work. The most advanced present-day time-marching techniques for solving the unsteady compressible three-dimensional Reynolds-averaged Navier-Stokes equations have their basis in simple time-marching techniques for solving the linear advection equation (1.15) - as well as on the general theory of hyperbolic conservation laws. By contrast, general solution techniques for the steady form of the governing fluid flow (Euler or Navier-Stokes) equations do not have such a sound infrastructure. It is well-known that the steady form of the equations are hyperbolic, parabolic or elliptic at a point on the flowfield according as to whether the local Mach number is greater than, equal to or less than one. Thus if zones of both supersonic and subsonic flow are present in the flowfield, the governing equations are 'mixed' and become quite awkward to tackle.

The simplest known mixed PDE is the Tricomi equation[14]:

$$yu_{xx} + u_{yy} = 0 \qquad \ldots (1.29)$$

(which changes from elliptic to hyperbolic or vice versa in any region containing a segment of the x-axis). But though it has been studied in considerable depth, methods for handling it can hardly be said to form a basis for more general mixed equations (though they certainly have contributed to an overall understanding of the problems involved).

Even in the case of flows which are completely subsonic, the elliptic system of equations which emerges has its own peculiar problems which are not easy to

---

[13] The reader will surely be aware that the steady form of the governing equations of fluid flow is hyperbolic only when the flow is supersonic.

[14] Named after F. Tricomi who first studied the equation in depth [Tricomi (1923)]. The books by Guderley (1962) and by Garabedian (1964) each devote a whole chapter to it.

simulate at a more fundamental level. The simplest known elliptic PDE, the Laplace equation:

$$u_{xx} + u_{yy} = 0 \qquad \qquad \dots (1.30)$$

bears little resemblance to the steady subsonic Euler equations - other than the fact that both are elliptic. It is not easy to see how an arbitrary method of treating the Laplace equation can be extended to deal with the steady subsonic Euler equations (even for incompressible flow); indeed, to the knowledge of the present author, none of the existing methods for treating the steady subsonic Euler equations, came into being as extensions of simpler elliptic solvers[15]. In fact the most that can be said is that methods for treating the steady subsonic Navier-Stokes equations have usually come into being as extensions of corresponding Euler solvers.

The best-known current method for treating the steady subsonic Euler or Navier-Stokes equations is probably the pressure-correction approach which had its origins in the early 1970s - about the same time as the introduction of the artificial time-marching approach in the context of CFD. Credit for its development is generally accorded to Patankar and Spalding (1972); the ideas behind the method are expounded in some depth in the book by Patankar (1980). Here let it merely be said that the very appellation 'pressure-correction' implies that the method can only be employed for problems where pressure plays a significant role, that is, the Euler or Navier-Stokes equations. In the context of general elliptic equations, the method becomes meaningless. Moreover, even in the context of the Euler or Navier-Stokes equations, pressure-correction methods become difficult to handle if zones of both supersonic and subsonic flow are present in the flowfield[16].

To put it another way, suppose we are required to solve an *arbitrary* system of partial differential equations. If the equations are time-dependent, it is natural (indeed virtually essential) to treat the problem by time-marching. However if the equations are not time-dependent, we have a choice. One method is to try to develop a special solver which will work only for the equations under consideration (such as pressure-correction for the Euler or Navier-Stokes equations). This special solver may need considerable effort (indeed research) to develop, because it cannot stand upon fundamental solution techniques for simpler equations. Moreover the solver may have to make use of a variety of strategies if the equations are mixed.

The other method - the method which forms the basis of the present work -

---

[15] For a survey of recent methods for treating the Euler equations, see for example Habashi (1985), Section 4, or Hirsch (1990), Part VI.

[16] These statements are not made with a view to criticism; indeed at low speeds the pressure-correction approach is generally preferable to time-marching (because low compressibility has the effect of slowing down a time-marching algorithm) and during the past decade so much research has been put in, that it is beginning to catch up with time-marching even in the high-speed domain.

is to artificially impose a time-dimension, and solve the resulting system of equations by time-marching towards a steady solution. Now a *uniform* mathematical treatment may be employed regardless of whether the original system of equations was elliptic, parabolic, hyperbolic or a mixture of all three - a uniform mathematical treatment whose 'roots' descend all the way to the simple one-dimensional linear advection equation. It is true that one may be able to find isolated examples where an artificial time-marching algorithm yields precisely the same algebraic steps as an iterative approach to the steady form of the equations [as we discovered in the case of Equations (1.6) and (1.7)], but such examples become increasingly difficult to find as the equations become more complex. If such identities can be found, they must be regarded as fortuitous.

It is now time to conclude this general introduction to the mathematical concept of time-marching and proceed with the specific problem of employing the time-marching approach for solving the two-dimensional Euler equations.

2 The Euler solver:
General layout of the program

In the introductory chapter of this work, it was stated that the biggest advantage of the time-marching approach was the fact that it rested on a solid body of fundamental work carried out on simple equations such as the linear and nonlinear advection equations (1.15) and (1.16) (page 7). It is appropriate to begin this chapter by considering a solution algorithm for these equations and showing how it can be logically extended to deal with the Euler equations.

Rather than take up (1.15) and (1.16) individually, we shall look at the general first–order hyperbolic equation (1.20) (page 10) - repeated here for convenience:

$$u_t + \{f(u)\}_x = 0 \qquad \ldots (2.1a)$$

or equivalently (with $g = f_u$)

$$u_t + g(u)u_x = 0 \qquad \ldots (2.1b)$$

For the problem to be properly posed, the function u must be specified at a given instant of time (which, without loss of generality, can be taken as $t = 0$) in either a finite or an infinite domain in x. If the domain is finite, boundary conditions will be required at the end points. In general these boundary conditions will remain unaltered at all subsequent instances of time.

All numerical solution algorithms for the finite problem must first involve the setting up of an appropriate grid - a trivial problem in one dimension. The initial values of u must then be read in (or, alternatively, calculated at each grid point from a given formula). The time-marching procedure may now commence. Referring to Figure 2.1a (page 17), the value of u at the point $(i,1)$ may be found *explicitly* in terms of the known values of u at the time–step $t_s = 0$, or *implicitly* in terms of other unknown values at the time–step $t_s = 1$ (as well as known values at $t_s = 0$). Depending upon the number of points that are fitted into the discretized equation, its order of accuracy can be increased as desired.

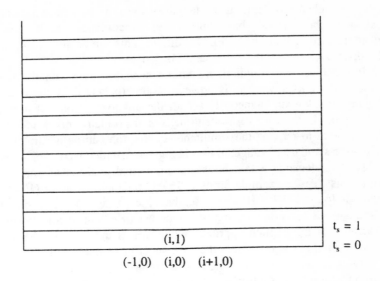

Figure 2.1a   **Time-steps for a one-dimensional problem**

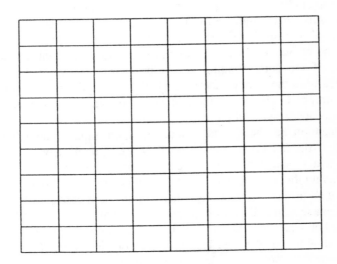

Figure 2.1b   **A rectangular Cartesian grid**
                    (the simplest case of gridding in two dimensions)

If the scheme is explicit, the length of the time-step Δt cannot be chosen at random, but must be limited in accordance with the so-called Courant-Friedrichs-Lewy (CFL) condition. This can be quite restrictive and will be discussed more fully in the next chapter.

If the scheme is implicit, no bound need be placed on the length of the time-step. This advantage is however counterbalanced by the disadvantage of having to solve simultaneously for all the unknown values of u at a particular time-step. In the case of a one-dimensional problem, this is not a particularly heavy price to pay, as the simultaneous solution can be accurately carried out using the standard tri-diagonal matrix algorithm[17] [details of which may be found in several elementary text books on numerical analysis or partial differential equations: see for example Smith (1965), Conte and De Boor (1972) or Phillips and Taylor (1973)]. However in the case of two or three dimensions, the simultaneous solution procedures cannot be carried out so conveniently. For this reason, it is the explicit approach which forms the basis of the present work, though no claim is made that they are superior to implicit methods.

But to return to the one-dimensional problem: for greater accuracy in the time discretization, the solution at a particular point on the first time step may be obtained in several stages: this is analogous to the Runge-Kutta schemes for solving ordinary differential equations. A one-stage scheme is at best first-order accurate, a two-stage scheme second-order accurate and so on.

Once the solution has been obtained at the first time-step, the same procedure may be applied for the computation of values at the next time-step and so on. The procedure may continue for as many time steps as desired. In general the question of 'convergence' to a steady state is irrelevant here because rarely if ever is the one-dimensional unsteady problem used as a tool for computing the solution of the corresponding steady problem. However if the process is continued until the steady state is reached[18], the result will be a linear function (i.e a straight line), the solution of "$u_x = 0$" satisfying the given boundary conditions at the two end points.

No mention has been made about the application of boundary conditions because in the one-dimensional case they are virtually trivial. In general the unknown u is specified at the two boundary points at all instances of time. Alternatively the derivative u_x may be specified at the two boundaries - or u may be specified at one and u_x at the other. In all these cases, only minor changes need be made in the discretization procedure. One other possibility is that of a periodic boundary condition in which the function u is assumed to be periodic, its period being equal to the length of the domain on which the initial data is specified and the solution is desired at future instances of time. This case does

---

[17] This assumes that the implicit approximation involves at most three neighbouring points at a particular time-step.

[18] Of course there is no guarantee that a steady state will be reached: the process may not converge.

not present much difficulty either: one need merely construct an additional grid point on the far side of either boundary and set the value of u at this grid point from its known 'counterpart' in the interior of the domain. The actual boundary position itself can then be treated as a normal grid point and the value of u at this point can be solved for in the usual way. The value of u at the other boundary point can then simply be set equal to it. An analogous procedure can be applied in two (or three) dimensions; precise details will be given in the next chapter.

This, then, is the 'program-sequence' of a solution algorithm for the one-dimensional equation (2.1). Although it seems comparatively simple, it is well to remember that considerable research has gone into questions such as stability, the elimination of spurious oscillations near shocks, and the relative merits/demerits of a large number of different schemes. However these need not concern us for the moment; our immediate objective is to see how the algorithm may be logically extended to two (or more) dimensions.

Incidentally, even in the context of one dimension, there is still scope for extension. The solution procedure just described is applicable only to the *scalar* equation (1.20), but the extension to the *vector* equation (1.21) does not present any great difficulty. True, there are now several unknowns instead of just one and each of these unknowns may depend nonlinearly on one or more of the others. But provided the manner of dependence is correctly known (as it must be if the system of equations is given), it should be possible to construct appropriate discretized relationships and solve these by time-marching. Other things being equal, the ratio of the total computational time taken for the vector solver to the time taken for the scalar solver should be of the same order as the number of unknowns involved.

While on the subject of the vector equation, there is one further point which should be mentioned. The scalar equation (1.20) is always hyperbolic, but the vector equation (1.21) is hyperbolic only if the eigenvalues are real and the eigenvectors linearly independent. Accordingly, if the method just described is to be applied to a vector equation, care should first be taken to ensure that these conditions are met with. If not the programmer will be committing the mathematical crime of using a solution technique totally alien to the physical nature of the problem and, needless to add, will not be able to arrive at a solution.

Let us now move on to the case of a scalar equation in two dimensions. The counterpart of Equation (2.1) is:

$$u_t + \{f(u)\}_x + \{g(u)\}_y = 0 \qquad \qquad \ldots (2.2a)$$

or equivalently,

$$u_t + f'(u)u_x + g'(u)u_y = 0 \qquad \qquad \ldots (2.2b)$$

For this problem to be properly posed, u must be specified at some fixed instant (say t = 0) and, if the domain is finite, appropriate conditions must be specified at all points on the boundary.

Right from the start we can see how much more complex a two-dimensional problem is in comparison to a problem in one dimension. In the latter case, the boundary degenerates to two distinct points, whereas in the former case, it is a curve which could be of any arbitrary shape and the conditions specified at one segment of this curve may be wholly different from those specified at another. Perhaps the simplest two-dimensional case is that in which the spatial domain is a rectangle as in Figure (2.1b) (page 17). But even here the complexities involved in relation to the one-dimensional case are considerably more than a single order-of-magnitude.

As with the one-dimensional problem, the solution algorithm must begin by setting up an appropriate grid. This is simple enough if the domain is rectangular as in Figure (2.1b), but for arbitrary domains can call for quite an exercise of skill[19]. Grid-generation will be discussed in a little more detail in the next chapter, but for the moment it is well to point out that it does not form an integral part of the present code which in fact assumes that a grid has been generated independently and is simply read in as part of the initial data.

Having generated the grid, the initial values of u must now be read in at each point of the grid just as in the one-dimensional case. If the problem is genuinely unsteady, these initial values must be *given*, but if we are merely using the unsteady form of the equation as a tool to arrive at the steady-state solution, the precise choice of the initial function is not so important as the final answer must be the same. Thus the initial values may be *guessed*, for example by interpolation from the known values on the boundaries. As with the grid, the precise method of choosing the initial guess is not an integral part of the present code. Instead it assumes that a guess has already been made and reads it as part of the initial data.

In the case of the one-dimensional problem, we were in a position to commence the time-marching algorithm as soon as all the initial data had been read in. In the case of two dimensions however there is generally some preliminary work to be done. Unless the grid is rectangular as in Figure 2.1b, the discretization of the governing equation (2.2) will involve coefficients which are functions of grid-cell areas, projective lengths on the coordinate axes etc. As these quantities are independent of the unknown u, it is advisable, for the purpose of computational economy, to compute them just once at the start of the program and store them in the memory of the computer - to be called upon as and when needed.

---

[19] In three dimensions, grid generation in arbitrary domains is not merely an exercise of skill; rather it is a problem at the research level.

The time-marching procedure may now commence. As with the one-dimensional problem, both explicit and implicit techniques may be adopted - but whereas the extension of the explicit approach is relatively straightforward, this is not the case with the implicit approach as we can no longer employ the efficient tri-diagonal matrix algorithm. For this reason, it is the explicit approach which is adopted here though, as mentioned on page 18, the time-step can be quite restrictive - all the more so in two dimensions. This will be discussed in more detail in the next chapter.

As in the one-dimensional case, the solution at a point on a particular time-step may be obtained in many stages, analogous to Runge Kutta methods for ODEs. The coefficients of these Runge-Kutta operators may be chosen so as to enhance stability, accuracy, etc.

The solution at a particular time step will not be complete until all the boundary conditions have been correctly imposed. Trivial though this may be in the context of one dimension, this is a subject which, from a two-dimensional point of view, merits careful study. We shall examine this question in some detail in the next chapter.

Having completed the solution on a particular time-step, the time marching procedure may be continued for as long as is desired. If the problem is genuinely unsteady, the solution at each time-step is an end in itself, but if the problem is 'pseudo-unsteady' the solution at each time-step is merely a means to an end, the end being the steady state which, hopefully, will be reached eventually. If the steady state is reached, we will have arrived at an answer to the partial differential equation:

$$\{f(u)\}_x + \{g(u)\}_y = 0 \qquad \qquad \ldots (2.3a)$$

or equivalently,

$$f'(u)u_x + g'(u)u_y = 0 \qquad \qquad \ldots (2.3b)$$

obtained from Equation (2.2) by elimination of the unsteady term.

Is there any guarantee that this steady state will be reached eventually? At the present time, this question cannot invariably be answered in the affirmative - there are simply too many factors to be taken into consideration. All that can be said is that the probability of reaching the steady state can be enhanced by adopting certain measures. In the first place, as has already been mentioned, the time-step should be chosen in accordance with the CFL criterion for stability. Secondly the scheme as a whole must contain a dissipative mechanism. This is automatically present in viscous computations, but not so when the governing equations are inviscid as in the present instance. It is therefore necessary to introduce artificial dissipation into the solution procedure; for the last decade, this has been an important subject of research. We will return to it in the next chapter.

The next stage in the hierarchy is the vector form of the two-dimensional problem:

$$U_t + F_x + G_y = 0 \qquad \ldots (2.4a)$$

or equivalently,

$$U_t + [F']U_x + [G']U_y = 0 \qquad \ldots (2.4b)$$

where U, F & G are vectors, and $[F']$ & $[G']$ are matrices with components:

$$f'_{ij} = \partial f_i/\partial u_j \quad ; \quad g'_{ij} = \partial g_i/\partial u_j. \qquad \ldots (2.5)$$

When dealing with this problem, the first step must be to ensure that it is indeed hyperbolic. This involves the computation of the eigenvalues and eigenvectors of the matrices $[F']$ and $[G']$, not always a trivial task. However once hyperbolicity is established, the main sequence of the solution procedure is essentially the same as the scalar case, each of the scalar variables which form the components of the vector U being dealt with individually at each grid point (or grid cell) on each time step, for as many time-steps as are desired or (in the case of pseudo-unsteady problems) until all the scalar variables converge to the steady state. It is true that the presence of several variables may complicate certain issues; for example one might have a negative effect on the convergence trend of another, or again the time-step restriction with regard to one variable may be quite different from that with regard to another. However these are simply 'finer points' which merit additional study: the essence of the solution algorithm remains unchanged.

It is now time to move from the general to the particular - that is to the Euler equations with which we are concerned in the present work. The two-dimensional Euler equations may be expressed vectorially in the form (2.4a) where

$$U = \begin{bmatrix} \rho \\ \rho u \\ \rho v \\ \rho e \end{bmatrix} \quad : \quad F = \begin{bmatrix} \rho u \\ \rho u^2 + p \\ \rho u v \\ \rho u h \end{bmatrix} \quad : \quad G = \begin{bmatrix} \rho v \\ \rho u v \\ \rho v^2 + p \\ \rho v h \end{bmatrix} \qquad \ldots (2.6)$$

The symbols which make up these vectors are well-known, but for the sake of completeness, let us write

ρ : density
u : velocity component in the x direction
v : velocity component in the y direction
p : pressure
e : energy
h : enthalpy.

Thus there appear to be six unknowns, but only four scalar p.d.e.s to solve for them. However it is possible to eliminate any two of these six unknowns by appropriate thermodynamic relationships which may be expressed algebraically. In fact the very nature of the governing equations suggests that we choose as our four fundamental variables the quantities ρ, ρu, ρv, ρe, which appear as the scalar components of the vector $\mathbf{U}$. All the other variables (even p and h) can now be expressed as algebraic functions of these four fundamental variables. In the next chapter, these relationships will be explicitly stated and it will be shown, in particular, how the scalar components of the vectors $\mathbf{F}$ and $\mathbf{G}$ can be expressed in terms of the fundamental variables. This will enable us to compute the components of the matrices $[\mathbf{F}']$ and $[\mathbf{G}']$, and hence to show that the unsteady Euler equations are indeed hyperbolic. For the moment, let us simply assume this to be true - or be content with the logical argument that any time-dependent problem must be hyperbolic (or at worst parabolic) because time-dependent variables can only depend on the past - not on the future[20].

We can now move on to the solution procedure. The first step is to set up a grid and an initial guess. As mentioned on page 20, this is not done in the program itself; it is up to the reader to generate a grid and an initial guess of his choosing and feed it into the program. Fortunately these tasks are fairly straightforward. The user does not need to take the trouble to generate an orthogonal grid; a simple H-grid will do the job[21]. As for the initial guess, all that need be done is to set the values of density, velocity etc from the known

---

[20] Suhas Patankar, in his much-read book *Numerical Heat Transfer and Fluid Flow* (1980) tries to explain the hyperbolicity of time-dependent equations in everyday language by saying "It is well known that tomorrow's events have no influence on what we do today". Though this seems convincing enough, the analogy is not really correct, because events which take place in our everyday lives are random and not subject to any mathematical law. To illustrate this point, we may consider the case of planetary movements which *are* subject to mathematical laws. By specifying a planetary configuration at a given instant, it is possible to investigate its past - thus providing a concrete example where the past may be computed from the future. This would be equally true with regard to our everyday lives if our actions were subject to mathematical laws. Fortunately (or unfortunately depending upon how one chooses to view it) they are not.

Nevertheless it should be possible to deduce by purely logical arguments that time-dependent problems are hyperbolic (or parabolic). The crucial factor is one-sided dependence; in the counterexample given earlier (that of investigating past configurations of planets), the unknown variables are computed *solely* in terms of known quantities at some future instant - not in terms of known quantities both in the future and in the past (two-sided dependence, characteristic of elliptic problems). We leave it to the reader to meditate more deeply on this subject if he so wishes.

[21] For the benefit of readers unfamiliar with this terminology, more details will be provided in the next chapter. Here let us just say that though an H-grid is simple to construct, it is not always a good choice because the nonorthogonalities which are generally present may lead to serious inaccuracies, or convergence difficulties, if the program is not well equipped for dealing with them. Thus a code which can yield satisfactory results on an H-grid is highly advantageous.

values at the boundaries. Of course it is quite possible that some variables may not be specified at any boundary; it is customary in fact to specify total pressure and total temperature rather than density and velocity. However if the problem is properly posed, it should be possible to obtain any desired variable by an appropriate algebraic formula. We will go into more details in the next chapter[22].

Since grid generation and setting of the initial guess are so simple, the reader may well ask why these tasks cannot be fitted into the program proper. The answer is that the ability to use a grid and an initial guess of one's own choosing provides the user with more flexibility. The program is designed to work on a wide variety of grids, so the augmentation of a particular grid generation routine would have the effect of restricting the range of solutions which the main solver is designed to provide. The same argument holds with regard to the initial guess.

Apart from the grid and the initial guess, various other data must be fed in so that the problem is properly defined. These include boundary conditions, coefficients of the Runge–Kutta scheme and of the artificial dissipation terms, relaxation parameters, printing criteria, convergence criteria, etc. The complete list of input data will be provided in the next chapter.

The second step in the solution procedure is the computation of all relevant quantities which will be made use of in the main solver but which, because they do not depend on any of the flow variables, need only be computed once at the start of the program and then stored in the memory of the computer. These include quantities such as cell-areas and projective lengths along the two Cartesian axes which will be required in the finite-volume approximation, as well as quantities such as radii of curvature at points on the blade surfaces, which will be required for the correct imposition of wall boundary conditions. The formulae used to compute these cell areas, projective lengths and radii of curvature will be explained in the next chapter. Note that whereas quantities such as radii of curvature only need be stored one-dimensionally, quantities such as cell areas will have to be stored two-dimensionally - at every cell in the flowfield.

This is an appropriate point to discuss, in a little detail, the question of computer storage. All CFD codes - indeed all problems associated with computers - come with a price: in terms of time and memory. In general if a CFD code is refined in order to yield better results, this can only be done at the expense of either time or memory (usually both). At present, research into means of reducing computational time (without the undue sacrifice of accuracy) is receiving a lot of attention - the reasons being purely economic: additional time means additional money, additional memory not necessarily so. Nevertheless computer memory is also a valuable commodity and any good computer

---

[22] For supersonic flow, all boundary conditions must be specified at inlet rather than outlet, so in effect the problem can be completely solved at the inlet by means of algebraic formulae alone. This is not true for subsonic flow where one condition must be specified at outlet. But regardless of whether the flow is subsonic or supersonic, our primary concern is to solve the problem in the main flowfield.

programmer will take care to use only as much storage as is genuinely necessary[23].

What would be the storage required for a time-marching solution to the Euler equations. First and foremost, one must store the grid. In two dimensions, this would require two 2-D arrays for each coordinate[24]. Then one must store the aforementioned projective lengths associated with the grid. In two dimensions this would require four arrays for the projective lengths and one for the cell areas.

Obviously we would require storage for the four fundamental variables ρ, ρu, ρv, ρe. Moreover it is also a great convenience to be able to store u, v & e independently for purposes of reading from the initial guess, imposition of boundary conditions, and printing of results. Associated variables such as pressure, enthalpy, temperature and speed of sound also require storage. So do mass fluxes in and out of each cell.

This is not all. Because the Runge-Kutta time-integration scheme involves several stages, the four fundamental variables ρ, ρu, ρv, ρe, would require temporary approximations ρ_t, ρu_t, ρv_t, ρe_t, which would be operated upon at each stage[25]. Then there are the quantities $\Delta\rho$, $\Delta\rho u$, $\Delta\rho v$, $\Delta\rho e$ representing the change in the fundamental variables from one time-step to the next. Then one must store the time-step itself; this is a function of the grid[26]. Finally there are the quantities relating to artificial viscosity: eight in two dimensions.

In the next chapter, we shall go into more details of each array - explaining the role they will play in the program as a whole - and mapping out the positions in the flow domain (i.e the computational grid) where they will require storage[27]. Let us now return to the solution procedure from the point where we left off - that is the computation of the projective lengths and cell areas.

---

[23] This is not to say that the programmer should be miserly in the allocation of computer storage. In several problems, additional storage arrays though not a strict necessity are nevertheless a great convenience. In these cases, the use of extra storage is certainly excusable if not actually desirable. In fact like money, computer storage should be used neither extravagantly nor parsimoniously but judiciously.

[24] If the grid is rectangular, two 1-D arrays would suffice.

[25] These temporary approximations should not be confused with time-derivatives.

[26] It is possible to make use of a fixed time-step throughout the grid, but this is unnecessarily wasteful of time, because in accordance with the CFL criterion for stability, it is possible to use much larger time-steps in some portions of the grid than in others. Naturally the larger the time-step, the quicker the steady state will be reached, so it is advisable to use the largest time-step that the CFL criterion will permit at any point of the flowfield. But this involves variation of the time-step in the flowfield.

[28] All the arrays mentioned in this list require two-dimensional storage, but not all require storage at boundaries, and some require storage at certain boundaries though not at others. It may be simpler to store every variable over the entire grid, but it is nevertheless instructive to depict the domain of each variable accurately.

Before the main time-marching procedure can commence, we must first 'set the scene' that is to say initialise all the variables that will be required in the discretized approximations to the continuity, momentum and energy equations. Thus far, we have only been considering the Euler equations as a whole; this is an appropriate moment to break it up into its four component equations:

$$\rho_t \;+\; (\rho u)_x \;+\; (\rho v)_y \;=\; 0 \qquad \dots (2.7a)$$
$$(\rho u)_t \;+\; (\rho u^2 + p)_x \;+\; (\rho u v)_y \;=\; 0 \qquad \dots (2.7b)$$
$$(\rho v)_t \;+\; (\rho u v)_x \;+\; (\rho v^2 + p)_y \;=\; 0 \qquad \dots (2.7c)$$
$$(\rho e)_t \;+\; (\rho u h)_x \;+\; (\rho v h)_y \;=\; 0. \qquad \dots (2.7d)$$

Equation (2.7a) expresses conservation of mass, (2.7b) & (2.7c) express conservation of momentum, and (2.7d) expresses conservation of energy. More briefly, they are known as the continuity, x-momentum, y-momentum, and energy equations.

In general, the initial guess will supply the values of density, velocities and pressure. From these, temperature can be obtained from the equation of state:

$$T \;=\; p/(\rho R), \qquad \dots (2.8)$$

where R is the gas constant (note that temperature is symbolized by a capital T to avoid confusion with 'time'). This leads us on to energy and enthalpy:

$$e \;=\; c_v T + \tfrac{1}{2}(u^2 + v^2) \qquad \dots (2.9a)$$
$$h \;=\; c_p T + \tfrac{1}{2}(u^2 + v^2) \qquad \dots (2.9b)$$

where c_v and c_p are the specific heats at constant volume and constant pressure respectively. Knowing u & v (from the initial guess) and e (from 2.9a), we can now initialize the four fundamental variables ρ, ρu, ρv and ρe.

We are now in a position to commence the time-marching process proper. First the temporary approximations ρ_t, ρu_t, ρv_t & ρe_t are set equal to ρ, ρu, ρv & ρe. The former will be updated at each stage of the Runge Kutta time-integration scheme, but the latter will remain fixed throughout the course of a particular time step. Next the value of the time-step must be determined at each individual cell. In accordance with the CFL criterion for stability, this will depend on the largest eigenvalues of the associated matrices [F'] & [G'] and, as we shall see in the next chapter, these are functions of the local speed of sound. Therefore we must first compute the local speed of sound c using the formula:

$$c \;=\; \{\gamma R T\}^{1/2}. \qquad \dots (2.10)$$

where γ is the ratio of specific heats: c_p/c_v. Having computed the local speed of sound at each grid cell, we may now evaluate the local time-step. The precise method of evaluation will be given in the next chapter.

26

Mass fluxes in and out of each cell face may be computed by appropriately multiplying velocity components and projective lengths. These fluxes are involved in all four of the governing equations (2.7 a,b,c,d). For example, in the case of the continuity equation (2.7a), the difference in the mass flux in and out of a particular cell, divided by the cell area and multiplied by the local time step yields the quantity $\Delta\rho$ by which the temporary density ρ_t will have to be updated at that particular cell. Similarly ρu_t, ρv_t & ρe_t may be updated by computing $\Delta\rho u$, $\Delta\rho v$ & $\Delta\rho e$ from the x-momentum, y-momentum and energy equations respectively. Specific formulae will be provided in the next chapter.

The statements made in the previous paragraph are not completely accurate because ρ_t, ρu_t, ρv_t & ρe_t are not updated solely from $\Delta\rho$, $\Delta\rho u$, $\Delta\rho e$ & $\Delta\rho e$, but also from the contribution of the dissipative terms which must be added for reasons of stability. Moreover in the initial stages of the Runge-Kutta time integration scheme, only a fractional contribution of the Δ and dissipative terms will be used in the updating process. More explicitly, the Δ and dissipative terms will have to be multiplied by suitable coefficients before the updating. These coefficients must be specified in advance, as part of the input to the program. The next chapter will provide some guidance as to the choice of suitable values.

At each time step, the temporary approximations ρ_t, ρu_t, ρv_t & ρe_t must be updated precisely as many times as the number of stages in the Runge-Kutta scheme. In general, a scheme of either three or four stages is employed - schemes of less than three stages tend to instability, schemes of more than four stages are unnecessarily wasteful of time. These assertions will be explained more fully in the next chapter.

An important point to remember is that at each stage of the Runge-Kutta scheme, ρ_t, ρu_t, ρv_t & ρe_t are updated by adding the Δ and dissipative terms (multiplied by the R-K coefficients) not to the old values of ρ_t, ρu_t etc but to the values of ρ, ρu etc.

After the last stage, all that is left to complete the time step is the imposition of the boundary conditions. Inlet, outlet, periodic and wall boundary conditions all require different methods of treatment. As usual we refer to the next chapter.

We conclude our operations on the time-step under consideration by setting the values of the fundamental variables ρ, ρu, ρv & ρe equal to the values of the temporary approximations ρ_t, ρu_t, ρv_t & ρe_t obtained at the last stage of the Runge-Kutta scheme.

This completes the cycle of operations at a given time-step. The cycle is now repeated at the next time-step ... and the next ... and the next ... until the steady state is reached. The attainment of the steady state may be ascertained with the aid of an appropriate convergence criterion. Once the steady state is reached, we will have arrived at the answer to our problem as stated on page 15: the computation of steady inviscid flow through a two-dimensional cascade.

It is now time to move on to the next chapter where we shall take a much more detailed look at some of the particular issues which were merely skimmed through in the present chapter.

3 The Euler solver:
An in-depth look at particular issues

While going through the previous chapter, the reader would have come across numerous repetitions of the phrase 'more details will be provided in the next chapter' (or words to that effect). To summarize, the main issues which the chapter skimmed over were: gridding, the associated geometric computations, the application of boundary conditions, the calculation of the time-step, the introduction of artificial dissipation, the selection of the initial guess, the proof of hyperbolocity of the unsteady Euler equations, the updating of the fundamental variables, the choice of the Runge-Kutta scheme to be used in the time-discretization and miscellaneous details associated with the programming of the code. Thus we have ten topics of discussion all of which have a role to play in the solution procedure as a whole. Let us now put these topics into some kind of order.

Because any realistic CFD problem[28] - regardless of the solution technique employed - must begin with the generation of a suitable grid, it is appropriate to begin with this subject as Topic 1 although, in order to provide the user greater flexibilty (see page 24, second paragraph), it does not form an integral part of the present code. Once the grid has been generated, the next logical topic of discussion is the geometric computation of grid-dependent quantities such as cell areas, projective lengths, etc. This is an instructive topic in its own right; such computations may be desirable in a wide variety of solution algorithms - not merely those involving time-marching or even computational fluid dynamics.

Having sorted out the problems with respect to gridding, we may move on to the

---

[28] By a realistic CFD problem, we mean a problem of practical interest, that is the computation of flow over an aircraft wing, turbomachinery blade, or more generally any configuration encountered in the 'real world'. Such configurations come in all kinds of shapes and sizes and the problem of generating a suitable grid may range from fairly easy to quite difficult.

time-marching approach proper. And, as we repeatedly emphasized in the previous chapter, before the time-marching approach can be applied to a system of partial differential equations, care must be taken to ensure that the system is indeed hyperbolic. So for our third topic of discussion, we shall express the components of the vectors [F] & [G] in the Euler equations in terms of the four fundamental variables ρ, ρu, ρv, ρe, and we shall use these to prove the hyperbolicity of the unsteady Euler equations.

Having proved hyperbolicity, we can safely commence operations knowing that we are not committing the mathematical crime of using a solution technique alien to the physical nature of the problem (page 19, fourth paragraph). And to commence operations, we must first set up the initial guess. As with grid-generation, this does not form an integral part of the present code thereby giving the user more flexibility. However some guidelines for setting up the initial guess will be presented here (Topic 4).

This may seem an appropriate place to list all the input data (point 9), but before doing so it is essential that we understand all the parameters involved, so we shall postpone this list to the section on general programming details.

The key to the time-marching approach is the computation of the quantities $\Delta\rho$, $\Delta\rho u$, $\Delta\rho v$ & $\Delta\rho e$ by which the four fundamental variables are to be updated at each time-step (after modification by artificial dissipation terms). However in order to compute these quantities, one must first set the time-step. As mentioned in the previous chapter, this is done locally at each individual cell. The time-step must not be too large or the method will be unstable; it must not be too small, as this would be wasteful of computational time. Determination of the appropriate time-step is a very important subject - intimately related to stability. Both these subjects will form Topic 5.

Topic 6 is the main algebraic discretization process of the spatially dependent terms in the Euler equations - from which the four quantities $\Delta\rho$, $\Delta\rho u$, $\Delta\rho v$, $\Delta\rho e$, are to be computed. These quantities must now be augmented with artificial dissipation terms - and this brings us on to Topic 7: artificial dissipation.

Once the Δ and dissipation terms have been obtained, the actual method of updating the fundamental variables will be carried out in several stages – analogous to Runge-Kutta methods for first-order differential equations. This will be Topic 8.

To complete the solution at a particular time-step, we must impose the four boundary conditions: inlet, outlet, periodic, wall - Topic 9.

Finally, having investigated all the essential topics, we can now list all the parameters that will be required as input, all the arrays that will have to be stored in the program, explain the role they play in the solution procedure, discuss the convergence criteria, the printing of results - in fact all general programming details, a preliminary to the actual listing of the Fortran program, which will follow in the next chapter.

Let us now summarise:

Topic 1. Gridding

Because grid-generation is not an integral part of the present code, we shall not dwell at much length on this subject. Given the geometry of a particular cascade, a user is expected to generate his own grid and feed it into the program. However for the benefit of readers who have little experience in this area, a few guidelines are suggested here.

Let us first have a look at three basic grid structures. As a model for our discussion, let us consider a 2-D blade section (or aerofoil section) composed of two circular arcs as in Figure 3.1a (page 31)[29]. It is our objective to compute the flow around this profile. One easy way to set up the grid would be to construct a set of 'quasi-streamlines' parallel to the profile and a set of vertical grid lines. Before and after the leading/trailing edge of the profile, the grid becomes rectangular. The characteristic H-shape of any particular section of the grid suggests the name H-grid: Figure 3.1b.

The biggest advantage of the H-grid is the simplicity with which it can be constructed. For a profile such as the double circular arc, it is generally quite satisfactory. In practice, however, many blade (or wing) cross-sections tend to have blunt leading edges as in Figure 3.2a (page 33). In such a case the use of an H-grid would lead to very skewed cells near the leading edge: Figure 3.2b. Moreover the quasi-streamlines will undergo a sharp change in direction in the region. Both these factors will lead to difficulties in the accurate representation of the governing equations in this region.

One approach to circumventing this difficulty is the construction of a so-called C-grid: Figure 3.2c. Here one set of grid lines wraps around the leading-edge of the profile in the shape of a C; the other set is approximately orthogonal. Although such a grid has the effect of obviating the aforementioned difficulties encountered by the H-grid, there is a price to be paid: there is no well-defined inlet boundary, so that

---

[29] Known as the double-circular-arc profile, this is a standard 'test-profile' on which to test the efficacy of a solution technique.

boundary conditions at inlet become awkward to impose.

A third type of grid structure – the so-called O-grid – can be generated for profiles having two blunt edges. Here the grid lines of one set are closed curves which wrap around the profile as a whole; the other set is approximately orthogonal. The reader can easily visualize such a grid without the necessity of having to depict a figure.

Thus far we have been considering gridding around a single profile, but – especially in the context of turbomachinery – we must also consider gridding between two blade profiles in a cascade. Once again the generation of an H-grid is fairly straightforward. It is true that the upper surface of the lower profile (*the suction surface*) and the lower surface of the upper profile (*the pressure surface*) will not be parallel to each other, but one need merely choose quasi-streamlines which gradually change their orientation from one surface to the other as in Figure 3.3a (page 35). The simplest way to do this is to construct the vertical grid lines first and then divide them evenly: this was the procedure followed in the construction of Figure 3.3a. Alternatively one may desire to concentrate the quasi-streamlines in the region near the two surfaces. This is only slightly more difficult: once again begin by constructing the vertical grid lines, but this time divide them in accordance with some formula that will concentrate grid points near the wall. Of course, the same formula must be used for each vertical grid line: Figure 3.3b

Here are two simple Fortran programs, the first of which generate the evenly-spaced grid shown in Figure 3.3a, and the second of which generates the weighted grids shown in Figure 3.3b.

```
-----------------------------------------------------------------------------------
C......PROGRAM 3.1:  GENERATION OF AN EVENLY SPACED H-GRID
       PARAMETER (M=40,N=20)
       DIMENSION X0(O:M),Y0(O:M),Y1(0:M),X(0:M,Y:N),Y(0:M,0:N)
       OPEN (1,FILE='PROFILE.DAT')
       DO 10 I = 0,M
       READ (1,*) X0(I),Y0(I),Y1(I)
       DY = (Y1(I)-Y0(I))/N
       DO 10 J = 0,N
       X(I,J) = X0(I)
 10    Y(I,J) = Y0(I) + DY*J
       STOP
       END
-----------------------------------------------------------------------------------
C......PROGRAM 3.2:  GENERATION OF A WEIGHTED H-GRID
       DIMENSION X0(0:M),Y0(0:M),Y1(0:M),X(0:M,0:N),Y(0:M,0:N)
       OPEN (1,FILE='PROFILE.DAT')
       READ *,W
       DO 20 I = 0,M
       READ (1,*) X0(I),Y0(I),Y1(I)
       DY = 0.5*(Y1(I)-Y0(I))*(W-1)/(W**(N/2)-1)
       X(I,0) = X0(I)
       Y(I,0) = Y0(I)
       DO 10 J = 1,N/2
```

32

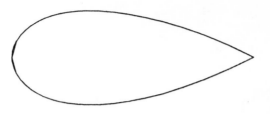

Figure 3.2a   **A round-nosed aerofoil**
                         (typical of many realistic problems)

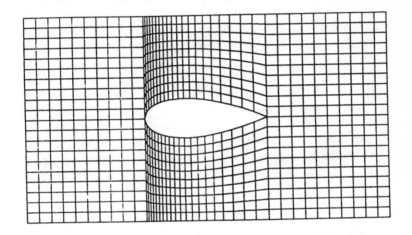

Figure 3.2b   **An H-grid about a round-nosed aerofoil**
                         (displaying skewed cells near the leading edge)

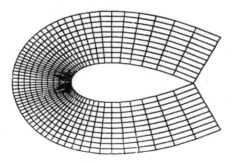

Figure 3.2c   **A C-grid about a round-nosed aerofoil**

```
      X(I,J) = X0(I)
10    Y(I,J) = Y(I,J-1) + DY*W**(J-1)
      DO 20 J = N/2+1,N
      X(I,J) = X0(I)
20    Y(I,J) = Y1(I) - (Y(I,N-J)-Y0(I))
      STOP
      END
```

A few comments: both programs read specified values of the grid points on the upper and lower boundaries and proceed to generate the rest of the grid. Note that whereas both X and Y coordinates are read at the lower boundary (X0,Y0), only the Y coordinate need be read at the upper boundary because the X-coordinate remains unaltered (by definition of an H-grid). Both boundaries incorporate the profile-shape itself as well as points upstream of the leading-edge and downstream of the trailing edge, which will be defined by periodic conditions in the main flow solver. In Figures 3.3, the profile is a double circular arc, but the two programs may be used for any other profile.

To run PROGRAM 3.2, the user must also feed in a weightage factor w. In general, this must be just a little greater than unity. The four grids depicted in Figure 3.3b use weightage factors progressively increasing from 1.1 to 1.4.[30] Incidentally the formulae used in this program are based on simple geometric progression. This is by no means the only way of concentrating the grid near the walls, but it is probably just as effective as any other - and simpler than most[31]. A reader interested in other formulae for refining grids near walls is referred to the paper by Roberts (1971).

The reader might wonder why the X-coordinates of the grid in the two programs are stored in a two-dimensional array when a one-dimensional array will suffice (in fact all values of X(I,J) for a particular value of I are precisely the same as X0(I)). The reason for this apparent wastage of array space is simply to keep the array format in parallel with that of the main Euler solver which is designed not merely for H-grids, but also for more general grids where both coordinates are genuinely two-dimensional. The Euler solver will read the grid -and it will expect it in the form of two 2-D arrays.

Figures 3.4 a,b,c (page 37) depict grids where the values of X(I,J) are genuinely two-dimensional. Note that the quasi-streamlines in Figures 3.4 a,b,c are precisely the same as in Figures 3.3 a,b,c respectively; in fact the three grids in Figure 3.4 have been obtained from their counterparts in Figure 3.3 by re-spacing the grid

----

[30] By a slight modification, the program can generate grids with different weightages at the two surfaces. It is left as an exercise.

[31] There is a lot to be said in favour of keeping programs simple - provided they do the work that is expected of them. To use a complicated mathematical tool when a simple one will do the job is something like using a bombastic word when a simple word will convey precisely the same meaning.

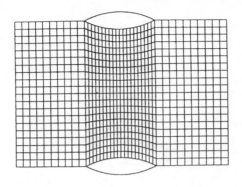

Figure 3.3a   **An H-grid through a double circular-arc cascade**

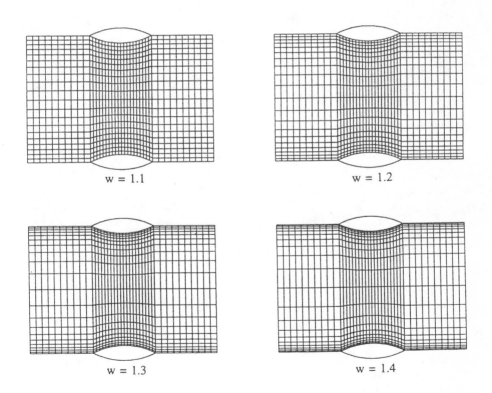

w = 1.1

w = 1.2

w = 1.3

w = 1.4

Figure 3.3b   **Weighted H-grids through the DCA cascade**

points along a particular quasi-streamline according to a given formula. Let us have a look at the Fortran program (page 36).

```
C.....PROGRAM 3.3:  GENERATION OF A SIMPLE QUASI-ORTHOGONAL GRID
      PARAMETER (M=40,N=20)
      DIMENSION X0(0:M),Y0(0:M),Y1(0:M),YH(0:M,0:N),FR0(0:M),FR(0:M),
     1    X(0:M,0:N),Y(0:M,0:N)
      OPEN (1,FILE='PROFILE.DAT')
      READ *,W
      DO 10 I = 0,M
      READ (1,*) X0(I),Y0(I),Y1(I)
      IF (W.EQ.1.0) DY = (Y1(I)-Y0(I))/N
      IF (W.NE.1.0) DY = 0.5*(Y1(I)-Y0(I))*(W-1)/(W**(N/2)-1)
      YH(I,0) = Y0(I)
      DO 10 J = 1,N/2
      IF (W.EQ.1.0) DYJ = DY
      IF (W.NE.1.0) DYJ = DY*W**(J-1)
   10 YH(I,J) = YH(I,J-1) + DYJ
      DO 20 I = 0,M
      X(I,0) = X0(I)
   20 Y(I,0) = Y0(I)
      DO 90 J = 1,N/2
      SQJ = SQRT(FLOAT(J))
      FR0(0) = 0.0
      DO 30 I = 1,M
   30 FR0(I) = FR0(I-1) + (X0(I)-X0(I-1)) + SQJ*ABS(Y0(I)-Y0(I-1))
      DO 40 I = 1,M
   40 FR0(I) = FR0(I)/FR0(M)
      DO 50 I = 1,M
   50 FR(I) = FR(I-1) + SQRT((X0(I)-X0(I-1))**2+(YH(I,J)-YH(I-1,J))**2)
      DO 60 I = 1,M
   60 FR(I) = FR(I)/FR(M)
      X(0,J) = X0(0)
      Y(0,J) = YH(0,J)
      II = 0
      DO 90 I = 1,M
   70 IF (FR(II).LT.FR0(I)) GO TO 80
      RATIO = (FR0(I)-FR(II-1))/(FR(II)-FR(II-1))
      X(I,J) = X0(II-1) + RATIO*(X0(II)-X0(II-1))
      Y(I,J) = YH(II-1,J) + RATIO*(YH(II,J)-YH(II-1,J))
      GO TO 90
   80 II = II+1
      GO TO 70
   90 CONTINUE
      DO 100 J = N/2+1,N
      DO 100 I = 0,M
      X(I,J) = X(I,N-J)
  100 Y(I,J) = Y1(I) - (Y(I,N-J)-Y0(I))
      STOP
      END
```

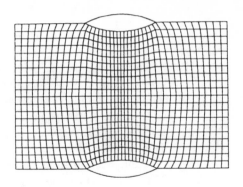

Figure 3.4a   **A quasi-orthogonal grid through the DCA cascade**

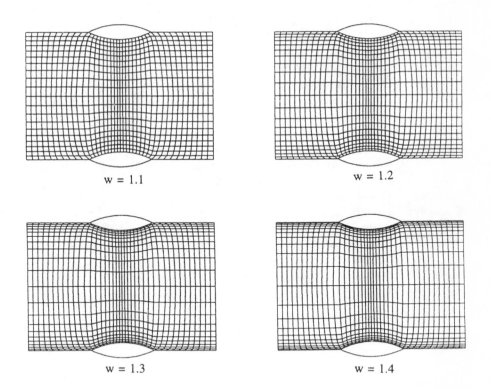

Figure 3.4b   **Weighted quasi-orthogonal grids through the DCA cascade**

Evidently PROGRAM 3.3 is more complicated than its two predecessors - and accordance with the philosophy outlined in footnote 31, page 34, our first question should be whether any advantage has been gained by the additional effort. The answer is yes - the grid cells are now somewhat more orthogonal, especially at the two surfaces. In general an orthogonal grid is a big asset, because even if the main flow solver is designed to operate on non-orthogonal grids, the presence of orthgonality will lead to a more accurate representation of the governing equations and a more accurate representation of boundary conditions. The price to be paid is the difficulty in generating an orthogonal grid in arbitrary domains. As aforementioned, PROGRAM 3.3 yields a grid that is somewhat more orthogonal than an H-grid, but this is all that can be said. A general program for generating an orthogonal grid is beyond the scope of the present manual.

PROGRAM 3.3 begins by generating an H-grid in precisely the same manner as in PROGRAM 3.1 and PROGRAM 3.2; indeed it combines the executable statements in both those programs in accordance with whether the weightage factor w (which must be read as input) is equal to or not equal to one. The Y-coordinates of the H-grid are stored in the array called YH; there is no need to store the X-coordinates as these are precisely the same as the wall grid-coordinates X0. Having obtained the quasi-streamlines (defined by the values of YH(I,J) for a fixed value of J), the remaining task is to re-space the grid points on each quasi-streamline so as to provide a more orthogonal structure. For this purpose two 'distribution functions' are evaluated, one relating to the grid-point spacing on the surface (FR0), the other relating to the grid point spacing on the quasi-streamline (FR). With the aid of these two functions, the grid points on the quasi-streamline are re-spaced. The precise choice of the two distribution functions are a matter of trial and error; what works well for one profile may not necessarily work well for another. However, with a little experience, a user should not find it difficult to select an appropriate distribution function.

We will not spend any more time on the subject of grid-generation; we hope the little that has been said will suffice to whet the reader's appetite for this important field. But before we pass on to Topic 2, we might add that the code which forms the subject of this work is so reliable than a simple H-grid usually suffices to give 'decent' results.

Topic 2. Geometric computations: projections, areas, etc

The grids depicted in the previous section may be considered (to use the jargon of set-theory) a union of disjoint cells. Each cell is an independent unit where the governing equations will be solved. According to the philosophy of the finite-volume methodology, on which the present code is based, the various conservation equations which collectively make up the unsteady Euler equations (or Navier-Stokes equations) must be precisely satisfied in each individual cell. Thus, according to the law of conservation of mass, the rate of change of density at a particular cell must be precisely equal to the difference between the mass flux entering that cell and the mass flux leaving that cell divided by the cell area. Similar statements hold for the laws of conservation of momentum and energy.

We see at once that the cell area is a very important factor in the computations and to save time, it must be computed in advance and stored for future use. What are the other grid-dependent quantities that will be required? To answer this question, let us see how the mass flux entering and leaving a particular cell is computed.

Consider Figure 3.5a (page 41). Five cells are depicted, the one in the centre (defined by the indices i,j) being the cell across which the various laws of conservation are to be satisfied. For the purpose of explanation it will suffice to consider the law of conservation of mass. Now the mass flux across the cell-face AC[32] may be split into two components - the flux through AG and the flux through GC. These fluxes may be obtained by multiplying the u & v velocity components[33] by the projections AG and GC respectively. The fluxes across the faces AB, BD and CD may be obtained in a similar manner.

It would seem therefore that for each cell, eight projective-lengths would require array storage. But this is an illusion. True, the finite-volume representation at each particular cell will have to make use of eight projective-lengths, but only half of these will actually require array storage. Thus for the index (i,j), it will suffice to store the four values AG, GC, CH & DH. Then in the course of the computations at (i,j), the values of EA & EB may be 'borrowed' from the index (i,j+1) and the values of BF & FD may be borrowed from the index (i+1,j)[34].

---

[32] Strictly we should say cell-edge, but in keeping with three-dimensional terminology, it has become common practice to use 'cell-face' in both two and three-dimensional problems. In fact the term 'cell-volume' is also often found in a two-dimensional context.

[33] These u & v components will be weighted averages of their values at the two cells which share the cell-face AC: (i,j) and (i-1,j).

[34] The reader might ask 'But what of the boundary cells where the indices (i+1) and (j+1) will not exist?'. The answer is that no projective-lengths will be needed here, because the flow variables will be determined through the application of boundary conditions.

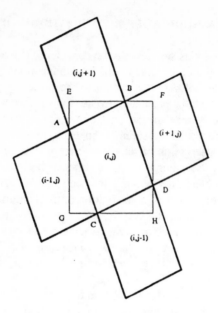

Figure 3.5a   **A grid cell and its neighbours (depicting projective lengths)**

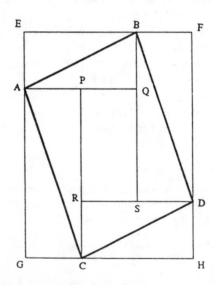

Figure 3.5b   **Determination of the area of the quadrilateral ABCD**

40

The actual evaluation of the projective lengths is straightforward[35]. The X and Y coordinates of the points A, B, C & D are known, and

EB = X(B)-X(A) : BF = X(D)-X(B) : GC = X(C)-X(A) : CH = X(D)-X(C)
$$\ldots (3.1a)$$
EA = Y(B)-Y(A) : FD = Y(B)-Y(D) : AG = Y(A)-Y(C) : DH = Y(D)-Y(C)
$$\ldots (3.1b)$$

What about the area of the cell as a whole? There are many ways in which it can be determined, but the method described here is quite 'elegant'.

Consider Figure 3.5b. ABCD, the cell whose area is to be determined, is an arbitrary quadrilateral, but EFGH and PQRS are both rectangles, their areas being:

Area (EFGH) = {X(D)-X(A)} x {Y(B)-Y(C)} $\ldots (3.2a)$
Area (PQRS) = {X(B)-X(C)} x {Y(A)-Y(D)}. $\ldots (3.2b)$

Furthermore:

Area (EFGH) = Area (ABCD)
                + Area (AEB) + Area (BFD) + Area (AGC) + Area (CHD) $\ldots (3.3a)$

Area (PQRS) = Area (ABCD)
                - Area (AQB) - Area (BSD) - Area (APC) - Area (CRD) $\ldots (3.3b)$

But triangles AEB, BFD, AGC & CHD are congruent to AQB, BSD, APC & CRD respectively. Thus Eqs (3.3a) and (3.3b) may be added to yield:

Area (ABCD) = 0.5 x [Area (EFGH) + Area (PQRS)]. $\ldots (3.4)$

Substitution of Eqs (3.2 a,b) gives the desired formula:

Area (ABCD)
   = 0.5 [{X(D)-X(A)}{Y(B)-Y(C)}+{X(B)-X(A)}{Y(A)-Y(D)}] $\ldots (3.5a)$

or, alternatively

Area (ABCD)
   = 0.5 [{X(D)-X(A)}{Y(B)-Y(C)}-{X(B)-X(A)}{Y(D)-Y(A)}]. $\ldots (3.5b)$

The second product in (3.5b) is the negative value of the area of the rectangle

---

[35] That is to say straightforward in the context of two dimensions. In three dimensions, these lengths become areas and their evaluation becomes somewhat more tricky.

PQRS and the reader might wonder what purpose is served by using this formula instead of the more straightforward (3.5a). From a purely computational point of view, the two formulae are exactly the same. The significance of formula (3.5b) is that the expression in the square brackets is the discrete representation of the differential form:

$$J = X_\xi Y_\eta - X_\eta Y_\xi, \qquad \qquad \dots (3.6)$$

where the suffixes ξ and η may be thought of as components of a curvilinear coordinate system aligned with the grid; ξ defines a measure along the quasi-streamlines or *streamwise grid lines* (i.e ξ increases monotonically and η is constant) and η defines a measure along the *transverse grid lines* (i.e η increases monotonically and ξ is constant). The differential form J in (3.6) is called the Jacobian; its significance will soon become clear.

Though a knowledge of the principles of curvilinear coordinate systems is not essential for an understanding of the present code, this is an appropriate point to make a few general remarks about this important subject. The major advantage of using a curvilinear coordinate system is that the so-called computational grid is effectively Cartesian[36]. Thus each cell in the ξ-η plane takes the shape of a rectangle and the discrete representation of the governing equations becomes straightforward. Referring to Figure 3.5a (page 41), there would be no need to compute eight projective-lengths AG, GC, etc. because four of these eight quantities would be equal to the four sides of the rectangle while the other four would be identically zero. And, of course, the computation of the cell areas would be equally straightforward.

Naturally there is a price to be paid. First and foremost, it should be remembered that the familiar form of the governing equations (Euler or Navier-Stokes) holds only for a Cartesian coordinate system. If the solution is to be found using a curvilinear coordinate methodology, then one must first transform the governing equations into the desired ξ-η coordinates. This transformation will involve the partial derivatives ξ_X, ξ_Y, η_X, η_Y, and the inverse transformation will involve the inverse partial derivatives X_ξ, Y_ξ, X_η, Y_η. These derivatives will have to be stored at each grid cell, though in practice only one of these two sets actually requires array storage; the four elements of the other can then be expressed in terms of analogous elements in the first. Suppose, for example, the set of four elements X_ξ, Y_ξ, X_η, Y_η, is stored, then the inverse derivatives may be obtained from the formulae:

$$\xi_X = Y_\eta / J \qquad \qquad \dots (3.7a)$$

---

[36] This sounds a bit confusing, but with a little reflection, the reader should have no difficulty in understanding the logic. In the X-Y plane, the coordinate system is Cartesian, but the grid is generally curvilinear. In the ξ-η plane, the coordinate system is curvilinear, but the grid becomes effectively Cartesian.

$$\xi_Y = -X_\eta / J \qquad \qquad \dots (3.7b)$$
$$\eta_X = -Y_\xi / J \qquad \qquad \dots (3.7c)$$
$$\eta_Y = X_\xi / J \qquad \qquad \dots (3.7d)$$

where J, the Jacobian[37] is defined by (3.6). Obviously J will also require storage at each grid cell. Eqs (3.7) are well-known, but like so many other well-known formulae, their proofs tend to be forgotten, so for the sake of completeness let us add that they can be proved by using the identities

$$X_\xi \xi_X + X_\eta \eta_X = X_X = 1 \qquad \qquad \dots (3.8a)$$
$$Y_\xi \xi_X + Y_\eta \eta_X = Y_X = 0 \qquad \qquad \dots (3.8b)$$
$$X_\xi \xi_Y + X_\eta \eta_Y = X_Y = 0 \qquad \qquad \dots (3.8c)$$
$$Y_\xi \xi_Y + Y_\eta \eta_Y = Y_Y = 1. \qquad \qquad \dots (3.8d)$$

Equations (3.8a) & (3.8b) can be solved for ξ_X and η_X to yield (3.7a) & (3.7c); Equations (3.8c) & (3.8d) can be solved for ξ_Y and η_Y to yield (3.7b) & (3.7d).

Now to return to the question of array storage. The use of curvilinear coordinates obviates the need to store projective lengths and cell areas, but in their place we would need to store the four derivatives X_ξ, X_η, Y_ξ, Y_η as well as the Jacobian J - making five arrays in all. We may recall that in the Cartesian coordinate system, we also have to store five arrays, four corresponding to the different kinds of projective lengths and one for the cell areas. But what is most interesting is that the projective lengths, as defined in equations (3.1), are simply discrete representations of the derivatives X_ξ, X_η, Y_ξ, Y_η, and the cell area, as defined in Equation (3.5b) is simply a discrete representation of the Jacobian J. In other words, whether we adopt a Cartesian coordinate framework, or a curvilinear coordinate framework, we are in practice following parallel paths - which is of course how it should be if both methods are to reach the same destination.

The advantages of the curvilinear coordinate formulation become a little more apparent in the context of three dimensions, because there is no additional difficulty in the computation of the derivatives X_ξ etc. (except that there are now 9 derivatives instead of 4) and the transformation formulae (3.7) as well as the Jacobian definition (3.6) can be logically extended to three dimensions. On the other hand in a Cartesian coordinate formulation, the computation of the projective lengths (now projective *areas*) and cell areas (now cell *volumes*) become considerably more complicated to work out (if we are to use the same geometric approach that was adopted in page 40). Nevertheless even in three dimensions, the two methods run in parallel. We will not pursue the three-dimensional solver in the present work, but hopefully it will be dealt with in a 'follow-up'.

Readers wishing to delve a little deeper into the subject of curvilinear coordinate

---

[37] The Jacobian is named after the 19th century German mathematician Karl Gustav Jacob Jacobi, who in his pioneering work De Formatione et Proprietatibus Determinantium called it a 'functional determinant'.

systems could refer to the CFD text book by Anderson, Tannehill and Pletcher (1984). See, in particular, Sections 5-6.2, where it is shown how the governing equations may be expressed in terms of curvilinear coordinates in such a manner so as to preserve the basic conservative form of the equations. More explicitly, the vector equation (2.4a) (page 22) is transformed to the form:

$$\mathbf{U}^*_t + \mathbf{F}^*_\xi + \mathbf{G}^*_\eta = 0 \qquad \qquad \dots (3.9)$$

where $\mathbf{U}^*$, $\mathbf{F}^*$ and $\mathbf{G}^*$ may be expressed in terms of $\mathbf{U}$, $\mathbf{F}$ and $\mathbf{G}$ by means of the Jacobian and the derivatives $\xi_X, \xi_Y, \eta_X, \eta_Y$. Incidentally the entire analysis is carried out in three dimensions, though only the two-dimensional analogue is described here.

Readers interested in studying curvilinear coordinates in a more practical setting could also refer to the book by the present author on acceleration of iterative schemes - Lobo (1992). There a potential flow-solver using a curvilinear coordinate methodology is described in considerable detail - with all the mathematical background and the complete Fortran listing.

Before moving on to Topic 3, one additional grid-dependent quantity must be examined - the radius of curvature of the profile at any grid point on the profile surface. The significance of the radius of curvature arises in the boundary condition for pressure: we shall prove, when we come to Topic 9, that the pressure gradient normal to the surface is given by

$$\partial p / \partial n = \rho (v_{tang})^2 / R_c \qquad \qquad \dots (3.10)$$

where n is the direction normal to the profile surface, v_{tang} is the velocity tangential to the surface, and R_c is the radius of curvature. It is worth mentioning that (3.10) is not the only method of expressing the boundary condition for the pressure-gradient - and the subject as a whole is still very much a research area. It will be discussed in more detail when we come to Topic 9, but here let it merely be said that Equation (3.10) is the boundary condition used in the present code[38] and, as such, it is necessary to evaluate the radius of curvature at each grid point on the profile surface. Incidentally although the radius of curvature serves no purpose in the program other than in its application to the pressure boundary condition, it is proper to discuss the subject here rather than in Topic 9, because just as is the case with the projective lengths and cell areas, it is one of those quantities which depend solely on the grid and not on the solution - and must therefore be evaluated at the

---

[38] By the present code is meant the code that is the subject of this book - not the original code developed by Cheng. As mentioned in the preface, the two are not identical, the present author having put in several changes, some for purposes of clarity, and some others which improved in some way upon the original at least in terms of computational economy. The use of the radius-of-curvature in the boundary condition for the pressure gradient is an example of the latter; Cheng's original method was rather cumbersome.

start of the program and stored in the memory of the computer, to be used as and when needed.

The method used here to compute the radius of curvature at a point on the profile consists of actually constructing a circle through this point and its two neighbouring grid points - and evaluating the radius of this circle. Obviously if the surface is flat, no circle can be drawn; the radius of curvature will be infinity. But this case can be ignored as the boundary condition (3.10) reduces to the trivial

$$\partial p/\partial n = 0. \qquad \qquad \dots (3.11)^{39}$$

Let us suppose that we wish to evaluate the radius of curvature at the point C on the profile surface in Figure 3.6a (page 47). Let B be the grid point immediately before C and A the grid point immediately after. Figure 3.6b provides an enlarged view of this section of the profile: a circle has been drawn through them. Let O be the centre of this circle and r its radius. Of course, at the moment, we do not know the value of r.

Let us make use of the conventional terminology from elementary geometry where a, b, c, represent the lengths of the sides of the triangle opposite the corners A, B, C, respectively. Let α and β be the angles subtended at O by the sides a and b respectively. Then using the familiar sine law of trigonometry, we have

$$\frac{a}{\sin\alpha} = \frac{r}{\sin(90-\alpha/2)} = \frac{r}{\cos\alpha/2}$$

or

$$\frac{a}{r} = \frac{\sin\alpha}{\cos\dfrac{\alpha}{2}} = \frac{2\sin\dfrac{\alpha}{2}\cos\dfrac{\alpha}{2}}{\cos\dfrac{\alpha}{2}} = 2\sin\dfrac{\alpha}{2}$$

Therefore

$$\alpha = 2 \sin^{-1} (a/2r) \qquad \dots (3.12a)$$

Similarly

$$\beta = 2 \sin^{-1} (b/2r) \qquad \dots (3.12b)$$

and

$$\alpha + \beta = 2 \sin^{-1} (c/2r). \qquad \dots (3.12c)$$

It is our objective to deduce a relationship between a, b, c and r; for this purpose,

---

[39] Equation (3.11) is sometimes used as an approximate boundary condition for the pressure-gradient; it is simple to apply and reasonably accurate if the radius of curvature is large.

we substitute Equations (3.12) into the trigonometric formula:

$$\sin\frac{\alpha+\beta}{2} = \sin\frac{\alpha}{2}\cos\frac{\beta}{2} + \sin\frac{\beta}{2}\cos\frac{\alpha}{2}$$

to obtain

$$\frac{c}{2r} = \frac{a}{2r}\sqrt{1-\frac{b^2}{4r^2}} + \frac{b}{2r}\sqrt{1-\frac{a^2}{4r^2}}$$

We must eliminate the radical sign. This requires two steps. First we square the left and right hand sides of the above equation to obtain:

$$\frac{c^2}{4r^2} = \frac{a^2}{4r^2}(1-\frac{b^2}{4r^2}) + \frac{ab}{2r^2}\sqrt{1-\frac{a^2}{4r^2}}\sqrt{1-\frac{b^2}{4r^2}} + \frac{b^2}{4r^2}(1-\frac{a^2}{4r^2})$$

There is now only one term containing radical signs. Taking it to one side and squaring once again, we have

$$\frac{a^2b^2}{4r^4}(1-\frac{a^2}{4r^2})(1-\frac{b^2}{4r^2})$$

$$= \frac{c^4}{16r^4} + \frac{a^4}{16r^4}(1-\frac{b^2}{4r^2})^2 + \frac{b^4}{16r^4}(1-\frac{a^2}{4r^2})^2$$

$$- \frac{a^2c^2}{8r^4}(1-\frac{b^2}{4r^2}) - \frac{b^2c^2}{8r^4}(1-\frac{a^2}{4r^2}) + \frac{a^2b^2}{8r^4}(1-\frac{a^2}{4r^2})(1-\frac{b^2}{4r^2})$$

Though this equation looks complicated, a great many of the terms cancel out including all terms containing r^8 in the denominator and all terms containing unequal powers of a & b. Simplifying and rearranging, we get

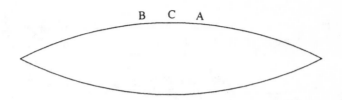

Figure 3.6a   **A profile surface:**
Computation of radius of curvature at the point C

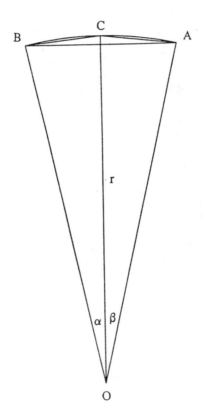

Figure 3.6b   **Enlarged view of section BCA:**
Computation of radius of curvature: r = OC

$$\frac{1}{16r^4}[2(a^2b^2+b^2c^2+c^2a^2)-a^4-b^4-c^4] \;=\; \frac{1}{16r^6}[a^2b^2c^2]$$

The expression in the square brackets on the left hand side is simply the product of the four terms (a+b+c), (a+b−c), (b+c−a) and (c+a−b). Therefore:

$$r \;=\; \frac{abc}{\sqrt{(a+b+c)(a+b-c)(b+c-a)(c+a-b)}} \qquad \ldots (3.13)$$

which is our desired formula.

Equation (3.13) is in fact a general formula for the radius of the circumcircle of a triangle whose sides are given by a, b and c. If any two of the three sides are equal, the triangle degenerates to a straight line and the radius becomes infinite. Therefore for most practical purposes, it is preferable to compute the inverse radius of curvature - which is really what is wanted for our present purpose - the pressure gradient boundary condition.

Here is a very simple subprogram that will compute the inverse radius of curvature, given the coordinates of three neighbouring points (X_1,Y_1), (X_2,Y_2), (X_3,Y_3). It may be used for general purposes - not merely for the present work.

C... SUBPROGRAM 3.4: COMPUTATION OF THE INVERSE RADIUS OF CURVATURE

```
FUNCTION RINV (X1,Y1,X2,Y2,X3,Y3)
A = SQRT((X2-X1)**2+(Y2-Y1)**2)
B = SQRT((X3-X2)**2+(Y3-Y2)**2)
C = SQRT((X3-X1)**2+(Y3-Y1)**2)
ABC  = A + B + C
AB   = A + B - C
BC   = B + C - A
CA   = C + A - B
RINV = SQRT(ABC*AB*BC*CA) / (A*B*C)
RETURN
END
```

There is an aesthetic simplicity about this little program; it is indeed remarkable that the cumbersome mathematical equations on the previous page lead up to such a simple algorithm. We will discover this again in our next section - the proof of the hyperbolicity of the unsteady Euler equations.

Topic 3. The unsteady Euler equations: proof of hyperbolicity

This section is of importance for two reasons. In the first place the substructure on which the code is built is being tested; if this substructure is not made of the appropriate material, the code will collapse. In the present case the substructure is the system of Euler equations and the appropriate material is hyperbolicity. So we must test for hyperbolicity. In the course of the test, we will have to evaluate the eigenvalues of relevant matrices - and this brings us to the second important factor: these eigenvalues define characteristic speeds and upon these characteristic speeds depend the time-step, the basic element of any time-marching code. Computation of the time-step is discussed in Topic 5, but now let us prove hyperbolicity.

For convenience, let us first re-write the vector form of the Euler equations:

$$\mathbf{U}_t + \mathbf{F}_x + \mathbf{G}_y = 0 \qquad \dots (3.15)$$

where

$$\mathbf{U} = \begin{bmatrix} \rho \\ \rho u \\ \rho v \\ \rho e \end{bmatrix} \quad : \quad \mathbf{F} = \begin{bmatrix} \rho u \\ \rho u^2 + p \\ \rho u v \\ \rho u h \end{bmatrix} \quad : \quad \mathbf{G} = \begin{bmatrix} \rho v \\ \rho u v \\ \rho v^2 + p \\ \rho v h \end{bmatrix} \qquad \dots (3.16)$$

The components of the vector $\mathbf{U}$ are our four fundamental variables. Let us refer to them as U_1, U_2, U_3 and U_4. Our first task is to express the components of the vectors $\mathbf{F}$ and $\mathbf{G}$ in terms of these four fundamental variables;

$$
\begin{aligned}
F_1 &= \rho u = U_2 && \dots (3.17a) \\
G_1 &= \rho v = U_3 && \dots (3.17b) \\
F_2 &= \rho u^2 + p && \\
     &= U_2^2/U_1 + U_1(\gamma\text{-}1)\,[U_4/U_1 - \tfrac{1}{2}\{U_2^2/U_1^2 + U_3^2/U_1^2\}] && \dots (3.17c) \\
G_2 &= \rho u v = U_2 U_3/U_1 && \dots (3.17d) \\
F_3 &= \rho u v = U_2 U_3/U_1 && \dots (3.17e) \\
G_3 &= \rho v^2 + p && \\
     &= U_3^2/U_1 + U_1(\gamma\text{-}1)\,[U_4/U_1 - \tfrac{1}{2}\{U_2^2/U_1^2 + U_3^2/U_1^2\}] && \dots (3.17f) \\
F_4 &= \rho u h = U_2 U_4/U_1 + U_2(\gamma\text{-}1)\,[U_4/U_1 - \tfrac{1}{2}\{U_2^2/U_1^2 + U_3^2/U_1^2\}] && \dots (3.17g) \\
G_4 &= \rho v h = U_3 U_4/U_1 + U_3(\gamma\text{-}1)\,[U_4/U_1 - \tfrac{1}{2}\{U_2^2/U_1^2 + U_3^2/U_1^2\}] && \dots (3.17h)
\end{aligned}
$$

Equations (3.17 a,b,d,e) are trivial; the other four make use of the formulae:

$$p = \rho(\gamma\text{-}1)\,[e - \tfrac{1}{2}(u^2+v^2)] \qquad \dots (3.18)$$
$$h = e + p/\rho. \qquad \dots (3.19)$$

Our next task is to compute the partial derivatives of each of these components

49

with respect to the four fundamental variables. These partial derivatives will form the components of the matrices [**F'**] and [**G'**] - see Equations (2.4b) and (2.5) (page 22) - whose eigenvalues have to be determined.

For reasons of space, the detailed derivations will be carried out only for the matrix [**F'**]:

$$\left[\frac{\partial F_1}{\partial U_1} = 0\right] : \left[\frac{\partial F_1}{\partial U_2} = 1\right] : \left[\frac{\partial F_1}{\partial U_3} = 0\right] : \left[\frac{\partial F_1}{\partial U_4} = 0\right]$$

$$\left[\frac{\partial F_2}{\partial U_1} = \frac{\gamma-3}{2}\frac{U_2^2}{U_1^2} + \frac{\gamma-1}{2}\frac{U_3^2}{U_1^2} = \frac{\gamma-3}{2}u^2 + \frac{\gamma-1}{2}v^2\right]$$

$$\left[\frac{\partial F_2}{\partial U_2} = (3-\gamma)\frac{U_2}{U_1} = (3-\gamma)u\right] : \left[\frac{\partial F_2}{\partial U_3} = (1-\gamma)\frac{U_3}{U_1} = (1-\gamma)v\right] : \left[\frac{\partial F_2}{\partial U_4} = (\gamma-1)\right]$$

$$\left[\frac{\partial F_3}{\partial U_1} = -\frac{U_2 U_3}{U_1^2} = -uv\right] : \left[\frac{\partial F_3}{\partial U_2} = \frac{U_3}{U_1} = v\right] : \left[\frac{\partial F_3}{\partial U_3} = \frac{U_2}{U_1} = u\right] : \left[\frac{\partial F_3}{\partial U_4} = 0\right]$$

$$\left[\frac{\partial F_4}{\partial U_1} = -\gamma\frac{U_2 U_4}{U_1^2} + (\gamma-1)\frac{U_2}{U_1}\left(\frac{U_2^2}{U_1^2} + \frac{U_3^2}{U_1^2}\right) = -\gamma ue + (\gamma-1)u(u^2 + v^2)\right]$$

$$\left[\frac{\partial F_4}{\partial U_2} = \gamma\frac{U_4}{U_1} - \frac{\gamma-1}{2}\left(\frac{3U_2^2}{U_1^2} + \frac{U_3^2}{U_1^2}\right) = \gamma e - \frac{\gamma-1}{2}(3u^2 + v^2)\right]$$

$$\left[\frac{\partial F_4}{\partial U_3} = (1-\gamma)\frac{U_2 U_3}{U_1^2} = (1-\gamma)uv\right] : \left[\frac{\partial F_4}{\partial U_4} = \gamma\frac{U_2}{U_1} = \gamma u\right]$$

Therefore

$$
[F'] \; = \;
\begin{bmatrix}
0 & 1 & 0 & 0 \\
\frac{\gamma-3}{2}u^2+\frac{\gamma-1}{2}v^2 & (3-\gamma)u & (1-\gamma)v & (\gamma-1) \\
-uv & v & u & 0 \\
-\gamma ue+(\gamma-1)u(u^2+v^2) & \gamma e-\frac{\gamma-1}{2}(3u^2+v^2) & (1-\gamma)uv & \gamma u
\end{bmatrix}
$$

For hyperbolicity, it is necessary to show that all the eigenvalues of this matrix are real. These *eigenvalues*[40] are defined by the solution of the matrix equation

$$[F'] \, X \; = \; \lambda \, X \qquad\qquad \dots (3.20)$$

where X is a vector (*eigenvector*) having the same number of elements as the vector F. Equivalently, they may be defined by the solution of the determinant equation:

$$\text{Det } [F'-\lambda I] \; = \; 0 \qquad\qquad \dots (3.21)$$

where I is the identity matrix. Now the determinant in (3.21) is

$$
\begin{vmatrix}
-\lambda & 1 & 0 & 0 \\
\frac{\gamma-3}{2}u^2+\frac{\gamma-1}{2}v^2 & (3-\gamma)u-\lambda & (1-\gamma)v & (\gamma-1) \\
-uv & v & u-\lambda & 0 \\
-\gamma ue+(\gamma-1)u(u^2+v^2) & \gamma e-\frac{\gamma-1}{2}(3u^2+v^2) & (1-\gamma)uv & \gamma u-\lambda
\end{vmatrix}
$$

To attempt to evaluate this determinant by a direct approach could prove quite cumbersome. Instead we shall first try to simplify it by making use of the following well known property of determinants:

The value of a determinant remains unaltered if the elements of one row (or column) are altered by adding to them any constant multiple of the corresponding elements in any other row (or column).

---

[40] The word *eigenvalue* derives from German, the prefix *eigen* meaning 'proper'.

Thus by multiplying column 2 by 'u' and adding it to column 1, we get:

$$\begin{vmatrix} u-\lambda & 1 & 0 & 0 \\ \dfrac{3-\gamma}{2}u^2+\dfrac{\gamma-1}{2}v^2-\lambda u & (3-\gamma)u-\lambda & (1-\gamma)v & (\gamma-1) \\ 0 & v & u-\lambda & 0 \\ \dfrac{\gamma-1}{2}u(v^2-u^2) & \gamma e-\dfrac{\gamma-1}{2}(3u^2+v^2) & (1-\gamma)uv & \gamma u-\lambda \end{vmatrix}$$

Next we multiply row 2 by 'u' and add it to row 4 to get:

$$\begin{vmatrix} u-\lambda & 1 & 0 & 0 \\ \dfrac{3-\gamma}{2}u^2+\dfrac{\gamma-1}{2}v^2-\lambda u & (3-\gamma)u-\lambda & (1-\gamma)v & (\gamma-1) \\ 0 & v & u-\lambda & 0 \\ u^2(\lambda-u) & \gamma e-\dfrac{3+\gamma}{2}u^2-\dfrac{\gamma-1}{2}v^2+\lambda u & 0 & u-\lambda \end{vmatrix}$$

Thirdly we multiply row 1 by 'u^2' and add it to row 4 to get

$$\begin{vmatrix} u-\lambda & 1 & 0 & 0 \\ \dfrac{3-\gamma}{2}u^2+\dfrac{\gamma-1}{2}v^2-\lambda u & (3-\gamma)u-\lambda & (1-\gamma)v & (\gamma-1) \\ 0 & v & u-\lambda & 0 \\ 0 & \gamma e-\dfrac{1+\gamma}{2}u^2-\dfrac{\gamma-1}{2}v^2+\lambda u & 0 & u-\lambda \end{vmatrix}$$

It is now time to introduce into the determinant the speed of sound c. Making use of the thermodynamic formula:

$$c^2 = \gamma p/\rho = \gamma(\gamma-1)[e-\tfrac{1}{2}(u^2+v^2)] \qquad \ldots (3.22)$$

the element in the fourth row, second column, becomes:

$$c^2/(\gamma-1) + \tfrac{1}{2}(v^2-u^2) + \lambda u.$$

The determinant has now been simplified to the extent where it is relatively easy to evaluate by the direct process of multiplying components. In fact if D_{ij} denotes the component in the ith row and j^{th} column, the determinant is simply

$$D_{11} \times [D_{22}D_{33}D_{44} - D_{23}D_{32}D_{44} - D_{24}D_{33}D_{42}] - D_{12}D_{21}D_{33}D_{44}$$

i.e.

$$(u-\lambda)[\{(3-\gamma)u-\lambda\}(u-\lambda)^2 + (\gamma-1)v^2(u-\lambda) + (\gamma-1)(\lambda-u)\{c^2/(\gamma-1)+\tfrac{1}{2}(v^2-u^2)+\lambda u\}]$$
$$- [u^2+\tfrac{1}{2}(\gamma-1)(v^2-u^2)-\lambda u](u-\lambda)^2.$$

We immediately see that $\lambda = u$ is a double root Equation (3.21), thereby providing us with two of the four eigenvalues. Both are real: so far so good. Now on dividing by $(u-\lambda)^2$:

$$[\lambda^2-(4-\gamma)\lambda u+(3-\gamma)u^2] + [(\gamma-1)v^2] + [-c^2-\tfrac{1}{2}(\gamma-1)v^2+\tfrac{1}{2}(\gamma-1)u^2+\lambda u]$$
$$+ [-u^2-\tfrac{1}{2}(\gamma-1)v^2+\tfrac{1}{2}(\gamma-1)u^2+\lambda u].$$

All the terms involving v (underlined) cancel out and the reader may easily verify that all the terms involving γ also cancel out. We are left with the simple equation:

$$\lambda^2 - 2\lambda u + u^2 - c^2 = (\lambda-u)^2 - c^2 = 0, \qquad \ldots (3.23)$$

the roots of which are $\lambda = u+c$ and $\lambda = u-c$. Thus the four eigenvalues are:

$$\lambda_1 = u \ : \ \lambda_2 = u \ : \ \lambda_3 = u+c \ : \ \lambda_4 = u-c. \qquad \ldots (3.24)$$

This is an appropriate point to recall the comment made in the previous section (page 48) in connection with the derivation of a formula for the radius of curvature. Despite the complexities in the algebraic process of derivation, the eventual formula is both simple and aesthetically pleasing. Precisely the same comment may be made in connection with the derivation of the eigenvalues of the matrix [F'. The answer, provided by (3.24), is indeed simple and aesthetically pleasing. And in a similar manner we can also show that the eigenvalues of the matrix [G'ɔare:

$$\lambda_1 = v \ : \ \lambda_2 = v \ : \ \lambda_3 = v+c \ : \ \lambda_4 = v-c. \qquad \ldots (3.25)$$

But we must not forget our main objective - which is not to show that the eigenvalues are aesthetically pleasing, but to prove that the unsteady Euler equations are hyperbolic. A sufficient condition for this is that all the eigenvalues of the matrices [F'] and [G'] are real and distinct. We have shown that they are all real, but unfortunately they are not all distinct. This does not imply that the equations are not hyperbolic. It simply means that we still have some work cut out.

Another sufficient condition for the hyperbolicity of the system of equations (3.15) is that the eigenvectors of the matrices [F'] and [G'] are linearly independent. This is a rather more general condition than the one previously stated, because from

53

a well-known theorem in linear algebra[41] the existence of a distinct set of eigenvalues automatically implies the existence of a linearly independent set of eigenvectors. However even if the eigenvalues are not all distinct, it may be possible to set up a linearly independent set of eigenvectors. Let us show how this can be done in the case of the Euler equations.

We can obtain a set of algebraic relationships between the four components x_1, x_2, x_3 & x_4 of the eigenvector X by substituting the elements of the matrix $[F']$ into the matrix equation (3.20). There are four equations altogether, but we need consider only three: the fourth will be automatically satisfied because of the vanishing determinant. The first equation is the simplest; it yields:

$$x_2 = \lambda x_1 \qquad \ldots (3.26a)$$

and the third is only slightly less trivial:

$$-uvx_1 + \lambda vx_1 + ux_3 = \lambda x_3 \qquad \ldots (3.26b)$$

i.e
$$(\lambda - u)x_3 = v(\lambda - u)x_1 \qquad \ldots (3.26c)$$

or if $\lambda \neq u$:
$$x_3 = vx_1. \qquad \ldots (3.26d)$$

We may now start the process of building up the eigenvectors. We can start by choosing x_1 fairly arbitrarily - the only real distinction is between a zero and a non-zero value, as different non-zero values would simply lead to eigenvectors that are scalar multiples of each other. We shall therefore only consider the cases $x_1 = 0$ & $x_1 = 1$. From Equation (3.26a), if $x_1 = 0$, then $x_2 = 0$, and if $\lambda \neq u$, then from (3.26d), $x_3 = 0$. However this will lead to the trivial solution in which all four components are zero, so except for the case when $\lambda = u$, we can dismiss the choice $x_1 = 0$. Thus for our four eigenvalues given by (3.24), we can build up four eigenvectors based on the first components $x_1 = 0$, $x_1 = 1$, $x_1 = 1$, and $x_1 = 1$ respectively. They would look like this:

$$X_1 = \begin{bmatrix} 0 \\ 0 \\ ? \\ ? \end{bmatrix} : \quad X_2 = \begin{bmatrix} 1 \\ u \\ v \\ ? \end{bmatrix} : \quad X_3 = \begin{bmatrix} 1 \\ u+c \\ v \\ ? \end{bmatrix} : \quad X_4 = \begin{bmatrix} 1 \\ u-c \\ v \\ ? \end{bmatrix} \qquad \ldots (3.27)$$

The question marks represent components that have yet to be filled. This is easy. Direct substitution in either the second or the fourth equation of (3.20) will show that x_4 must be equal to q $[=\frac{1}{2}(u^2+v^2)]$, $h+uc$ and $h-uc$ for the eigenvectors X_2, X_3 and X_4 respectively. In the case of X_1, any non-zero value may be selected for x_3 (a zero value would lead to the trivial solution). Choosing $x_3 = 1$, we obtain $x_4 = v$.

We can now write the complete eigenvector set:

---

[41] See, for example, Kreyszig (1988), Section 7.15.

$$
x_1 = \begin{bmatrix} 0 \\ 0 \\ 1 \\ v \end{bmatrix} \quad : \quad x_2 = \begin{bmatrix} 1 \\ u \\ v \\ q \end{bmatrix} \quad : \quad x_3 = \begin{bmatrix} 1 \\ u+c \\ v \\ h+uc \end{bmatrix} \quad : \quad x_4 = \begin{bmatrix} 1 \\ u-c \\ v \\ h-uc \end{bmatrix} \qquad \text{... (3.28)}
$$

One task remains: to prove that this set is linearly independent. This really should be obvious from the manner in which it was constructed, but for the sake of completeness let us provide a rigorous proof.

If any linear dependence exists between the set of vectors (3.28), then there must exist a set of scalars a_1, a_2, a_3 and a_4 such that

$$
a_1 X_1 + a_2 X_2 + a_3 X_3 + a_4 X_4 = 0. \qquad \text{... (3.29)}
$$

This results in the four equations:

$$
\begin{aligned}
a_2 + a_3 + a_4 &= 0 & \text{... (3.30a)} \\
u a_2 + (u+c) a_3 + (u-c) a_4 &= 0 & \text{... (3.30b)} \\
a_1 + v(a_2 + a_3 + a_4) &= 0 & \text{... (3.30c)} \\
v a_1 + q a_2 + (h+uc) a_3 + (h-uc) a_4 &= 0. & \text{... (3.30d)}
\end{aligned}
$$

From (3.30 a,b):	$a_3 = a_4$	... (3.30e)
From (3.30 a,c):	$a_1 = 0$	... (3.30f)
From (3.30 a,e):	$a_2 + 2a_3 = 0$	... (3.30g)
From (3.30 d,e,f):	$q a_2 + 2 h a_3 = 0$	... (3.30h)
From (3.30 g,h):	$q = h.$	... (3.30i)

The last equation is impossible; in fact, from (3.18) & (3.19), page 49:

$$
h = e + (\gamma - 1)(e - q). \qquad \text{... (3.31)}
$$

Thus there cannot exist any set of constants satisfying Equation (3.29) and the proof of the linear independence of the eigenvectors (3.28) is complete.

We have gone through this detailed analysis only for the matrix $[\mathbf{F}']$. In order to complete the proof of hyperbolicity of the Euler equations, we ought to repeat it for the matrix $[\mathbf{G}']$, but as the process is entirely parallel, it may be safely left as an exercise to the reader. We can now move on to the time-marching approach proper, safe in the knowledge that we are treading on firm ground.

Topic 4. The initial guess

At the beginning of the previous section, we drew an analogy between the proof of the hyperbolicity of the unsteady Euler equations and the testing of the substructure of a building. The fact that the equations were found to be hyperbolic meant that the substructure was sound and the time-marching process could commence. Continuing with our analogy, we can now compare the actual process of time-marching to the actual construction of the building. Each floor of the building may be compared to a time step - the ground floor representing the initial guess. And just as the construction of any building commences at the ground floor, so also does the execution of any time-stepping algorithm commence with the initial guess.

In this architectural procedure, the user will be expected to design his own ground floor. And just as the actual architect of a building is bound by certain constraints (for example the outer walls may have to be set to a prescribed plan), so also is the user of a computer program bound by certain prescribed conditions. In the case of two-dimensional flow over a cascade (with which we are concerned in the present work), the conditions generally prescribed are the total pressure at inlet, the total temperature at inlet and a third condition either at inlet or outlet depending upon whether the freestream Mach number is supersonic or subsonic. These boundary conditions will be discussed in some depth in Topic 9, but for the moment we are concerned only with the role they have to play in the setting-up of the initial guess.

The simplest form of the initial guess is to set values of density, velocity and pressure that are constant throughout the flowfield[42]: which means that they can be set equal to their known values at inlet - if they are all known at inlet. This is true enough at supersonic freestream Mach numbers, because the third condition (apart from stagnation pressure P_0 and stagnation temperature T) completes the specification. Without loss of generality, we may assume that the third condition is the freestream Mach number M itself. Then the static pressure p and static temperature t may be obtained from the thermodynamic formulae:

$$ p = P_0 \left[\frac{1}{1 + \frac{\gamma-1}{2}M^2} \right]^{\frac{\gamma}{\gamma-1}} \quad ; \quad t = T_0 \left[\frac{1}{1 + \frac{\gamma-1}{2}M^2} \right] \qquad (3.32 \text{ a,b}) $$

where γ is the ratio of specific heats. The derivation of these formulae may be found in various text books on gasdynamics, for example Liepmann and Roshko (1956), Section 2.10, or Cambel and Jennings (1958), Sections 3.6, 3.7. Many similar formulae may be found in Table 2.5 in the book by Oswatitsch (1956).

Having obtained pressure and temperature, we may compute density from the equation of state:

---

[42] Strictly, the initial guess must be set up in terms of the four fundamental variables ρ, ρu, ρv, & ρe. But because the user will be more 'familiar' with pressure rather than energy, and to save him the trouble of multiplication, the program reads the initial values of ρ, u, v & p.

$$\rho = \frac{p}{Rt}, \qquad \dots (3.33)$$

the speed of sound from either of the two formulae:

$$c = \sqrt{\gamma \frac{p}{\rho}} \quad : \quad c = \sqrt{\gamma Rt} \qquad \dots (3.34)$$

and the total velocity by multiplying the Mach number with the speed of sound. For the purpose of the initial guess, we may assume that this total velocity is directed parallel to the mainstream flow and compute u and v accordingly.

What if the mainstream flow is subsonic? In this case one inlet condition will have to be dropped and in its place an outlet condition added[43]. In practice it is customary to specify the stagnation pressure and temperature at inlet (just as in supersonic flow) and the static pressure at outlet. Now at neither boundary is our information sufficient to obtain the other flow variables but we could operate at the outlet, if we assume that there is no change in stagnation pressure and temperature - reasonable enough for the purpose of an initial guess. Knowing P_0 and p, the Mach number may be obtained from the thermodynamic formula:

$$M = \sqrt{\frac{2}{\gamma-1}\left[\left(\frac{P_0}{p}\right)^{\frac{\gamma-1}{\gamma}} - 1\right]} \qquad \dots (3.35)$$

The other flow variables may now be obtained as before.

Obviously the method which has just been outlined is not the only way of setting up the initial guess: if it was, it would have been incorporated into the program, whereas in fact the program simply accepts whatever initial guess is fed in by the user. There are various methods by which the user may attempt to improve upon the above method, for example the velocity need not be kept constant throughout the flowfield, but could be varied in accordance with a one-dimensional mass flow relationship. This may have the effect of reducing the total number of time steps required by the main flow solver to reach a steady state - thereby also reducing the total computational time. In fact the method which has been described here is just about the simplest means of generating an initial guess. But the time-marching approach is so effective that a simple initial guess is all that is required.

It is now time to begin our ascent from the ground floor ...

---

[43] The reason why the specification of boundary conditions changes with the physical nature of the problem will be explained more fully when we come to Topic 9.

Topic 5. Stability and its effect on the time-step

Continuing our analogy with the construction of a building, we have now completed the ground floor and must move on to floor 1 ... then on to floor 2 ... and so on. Unfortunately we are not free to design these floors as we please; the design has to follow certain mathematical laws. By means of these mathematical laws, the design on each floor will relate - to a greater or lesser extent - to the design at the floor immediately below it; this relationship will depend partly on the distance between the two floors and partly on the height of the floors above ground level. Once a certain height has been reached, the design on each floor will have grown independent of its predecessor. At this stage, we will have attained an 'ideal' design and can stop the construction process.

Naturally great care must be taken in the building process as otherwise a critical point may be reached when the whole structure may collapse. What are the factors which may bring about this catastrophe? One obvious factor is the gap between floors. If it is too large, the structure may become unstable. On the other hand the altitude of our top floor is very high indeed and as our objective is simply to get there, we do not wish to spend an inordinately large amount of time building intermediary floors. Ideally we wish to build only just as many intermediary floors as will suffice to ensure that the construction as a whole is stable. Or, what amounts to the same thing, we wish to be able to impose a gap between floors that is as large as possible - but without endangering stability.

There are of course several incongruities in this analogy between a time-marching algorithm and the construction of a building. For example, each floor of a building must be 'horizontal', the gap between any two floors being constant at all locations, but the time-step (at least in an *artificial* time-marching algorithm) may vary at different points of the grid. Again, by increasing the number of supporting pillars, one would improve the stability of a building, but though, intuitively, the pillars may be likened to the points on the grid, the analogy fails because grid-refinement will not necessarily improve stability. Indeed paradoxically it puts an even more severe restraint on the size of the time step, because as we shall now show, stability considerations for explicit algorithms require that the time-steps relate to the mesh-size in direct proportion.

The stability of a time-marching algorithm is so basic that it can be studied in the context of the simple linear advection equation (1.15a) - repeated here for convenience:

$$u_t + \lambda u_x = 0 \qquad \qquad \ldots \text{(3.36a)}$$
$$u(x,0) = f(x). \qquad \qquad \ldots \text{(3.36b)}$$

The constant c in Equation (1.15a) has been changed to λ partly in order to avoid confusion with the speed of sound and partly in order to stress the parallel with the eigenvalues that were computed in Topic 3 of this chapter, which as we shall see play an analogous role in carrying information along characteristic curves. As mentioned in Chapter I, Equation (3.36) has a trivial analytic solution

$$u(x,t) = f(x-\lambda t) \qquad \ldots (3.37)$$

but its importance lies in the fact that it is the most basic of all test cases for numerical approaches to solving hyperbolic conservation laws. A numerical approximation to Equation (3.36) (and more general conservation laws) must involve both time and space discretization. Most schemes of practical interest make use of only two time-levels for the purpose of time-discretization[44], though this does not necessarily restrict the procedure to first-order accuracy, because the computation can be carried out over a number of stages just as in a Runge-Kutta method for solving ODEs. The Runge-Kutta integration process will be discussed in Topic 8, but for the moment let us assume that the time-discretization is carried out via a straightforward backward difference formula. Then, assuming the discretization is explicit, Equation (3.36a) may be approximated as:

$$u(i,j+1) = \Sigma c_m u(i+m,j) \qquad \ldots (3.38)$$

where m ranges over a certain interval and in general includes both negative and positive values. Equation (3.38) arises by combining the first-order approximation for the time-derivative:

$$u_t = [u(i,j+1)-u(i,j)] / \Delta t \qquad \ldots (3.39)$$

with a general approximation for the space-derivative:

$$u_x = [\Sigma a_m u(i+m,j)] / \Delta x, \qquad \ldots (3.40)$$

where it has been assumed for the sake of illustration that the grid-spacing in the x-direction is constant. Then the coefficients c_m in Equation (3.38) are given by

$$c_0 = 1 - a_0 (\lambda \Delta t/\Delta x) \qquad \ldots (3.41a)$$
$$c_m = -a_m (\lambda \Delta t/\Delta x) \text{ if } m \neq 0. \qquad \ldots (3.41b)$$

The quantity $\lambda \Delta t/\Delta x$, known as the Courant number[45], is of great importance in the context of stability analysis.

The standard method of analysing the stability of the scheme (3.38) (and also of

---

[44] The major disadvantage of schemes involving three or more time-levels is that they are difficult to get 'off the ground', because to start with we are only provided with data at a single time level.

[45] After Richard Courant (1888-1972), a German-born mathematician, best known for his work with David Hilbert; the pair wrote *Methods of Mathematical Physics*; he emigrated to USA following the rise of the Nazis and was a professor at New York University until his retirement in 1958; he founded a prestigious institute of applied mathematics in New York, named in his honour after his retirement; see footnote 11, page 11.

more general schemes) is the so-called Von Neumann method[46]. This method is discussed in virtually every text book on the numerical solution of partial differential equations - it rates a whole chapter in the two-volume CFD text-book by Hirsch (1988). Basically the idea is to associate with a given algorithm, an 'amplification factor' which indicates how the errors (which may arise due to discretization, round-off, etc.) will behave in the course of the solution procedure. If the absolute value of the amplification factor is less than unity, the errors will eventually die away; if not they may grow without bound and the algorithm will become unstable.

In the case of the algorithm (3.38), the associated amplification factor is

$$AF(\theta) = \Sigma c_m e^{im\theta} \qquad \qquad \ldots (3.42)$$

where i is the square root of -1. Thus stability of the algorithm (3.38) would require that (3.42) have an absolute value less than one for all values of the frequency θ. Observing, from (3.41), that the coefficients c_m are functions of the Courant number, it follows that the stability of the algorithm (3.38) will depend on the Courant number. If the Courant number exceeds a certain critical value, the algorithm will become unstable.

Evaluation of the critical value of the Courant number for algorithms of general order is a fairly difficult task - even in the case of the linear advection equation. The reader interested in pursuing this subject could refer to Strang (1962). The present author has also done considerable work in this area - which has been summarised in a Cranfield College of Aeronautics report - Lobo (1985). At the time, the author was unaware of the work of Strang and as it happened, some of the results proved were the same though the methods used were quite different.

However if the order of the algorithm is not too large, the procedure simplifies. The reader will probably be aware of the special case where the spatial derivative is discretized using a first order backward difference formula[47] (the standard text-book example for outlining the Von Neumann method). Here the algorithm (3.38) reduces to

$$u(i,j+1) = \sigma u(i-1,j) + (1-\sigma)u(i,j), \qquad \ldots (3.43)$$

where $\sigma = \lambda \Delta t/\Delta x$ is the Courant number. The amplification factor (3.42) becomes

---

[46] After John Von Neumann (1903-57), a Hungarian-born mathematician, who did extensive research both in pure and applied mathematics, though he is now best remembered as a pioneer in the development of the computer. He too emigrated to USA in the 1930s and worked as a consultant to the armed forces; it was during the War that he worked out his theories on stability, though for security reasons his work was not published - so unfortunately there is no 'classic paper' on Von Neumann's stability theory.

[47] This assumes that λ is positive. If λ is negative, a forward difference formula must be employed.

$$AF(\theta) = \sigma e^{-i\theta} + (1-\sigma) \qquad\qquad \ldots (3.44)$$

and it can easily be verified that for its absolute value to be less than unity (for all values of θ), the Courant number σ must lie between zero and one.[48].

The reader may have noted that in formula (3.43), u(i,j+1) is simply expressed as a weighted average of u(i-1,j) and u(i,j), the weightage depending on the value of the Courant number. Recalling that the exact solution of the linear advection equation is given by Equation (3.37), the value of u(i,j+1) will in fact be equal to u(i-σ,j) - Figure 3.7a (page 63). Indeed this provides us with another means of deriving (3.43), as well as the more general algorithm (3.38); in both formulae, if we replace the left hand side with u(i-σ,j), the right hand side is simply the interpolatory formula of appropriate order.

In Figure 3.7a, (i-σ,j) is depicted as lying *between* (i,j-1) and (i,j), but this is true only if σ lies between zero and one (the stable case). If σ does not lie between 0 & 1, u(i-σ,j) will lie outside these two limits, so that formula (3.43) becomes a matter of extrapolation rather than interpolation. It would seem therefore that extrapolation tends to be more unstable than interpolation, a fact that it is well known even in elementary numerical analysis, without having to resort to Von Neumann stability techniques.

To put it more expressively, the interval between (i,j-1) and (i,j) may be regarded as a kind of firm ground. The numerical solution at (i,j+1) is supported by a 'stalk' defined by the characteristic dx/dt = λ. If this stalk rests on firm ground, all is well, but if it springs from thin air, the solution will be unstable.

How does this philosophy influence the computation of the time step? Well assuming that the spatial grid is fixed, Δt must be chosen such that

$$\sigma = \lambda \Delta t / \Delta x < 1, \text{ that is, } \Delta t < \Delta x / \lambda.$$

Assuming that $\lambda > 0$, this is always possible, though the more refined the grid, the greater the restriction on the size of the time step.

What if $\lambda < 0$? In that case, the Courant number will be negative and the scheme (3.43) is always unstable. To put it another way, the stalk that supports the solution at the point (i,j+1) springs from the right hand side of the grid point (i,j) and we must therefore choose as our 'firm ground' an interval to the right; in other words employ a forward difference scheme. The forward difference analogue of (3.43) is

$$u(i,j+1) = (1+\sigma)u(i,j) - \sigma u(i+1,j) \qquad\qquad \ldots (3.45)$$

for which the amplification factor is

$$AF(\theta) = (1+\sigma) - \sigma e^{i\theta} \qquad\qquad \ldots (3.46)$$

---

[48] The values of the amplification factor will in fact describe a circle centered at $(1-\sigma,0)$ with radius σ: see Hirsch (1988), Figure 8.1.2.

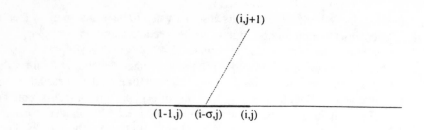

Figure 3.7a   **Propagation of the 1-D solution from (i-σ,j) to (i,j+1)**

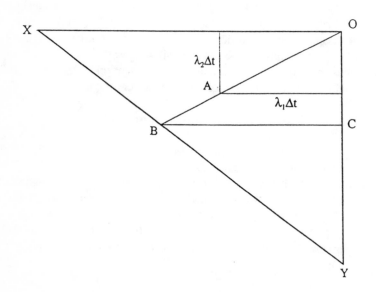

Figure 3.7b   **Analysis of the CFL condition in two dimensions**

and it may easily be verified that for stability, σ must satisfy $-1 < \sigma < 0$. The time-step Δt can now be computed such that $\Delta t < -\Delta x/\lambda$.

Now let us move a step foward in the hierarchy, that is to say consider the general first order system (1.21a) (page 10). Here we have not one but several characteristics - defined by the eigenvalues of the matrix $[\mathbf{F}\,']$. Each of these characteristics may be regarded as a stalk which supports the (vector) solution. And if the solution procedure is to be stable, each of these stalks must rest on firm ground. To return to PDE jargon, suppose $'dx/dt = \lambda_+'$ and $'dx/dt = \lambda_-'$ denote the extreme characteristics[49], then the interval between the points at which they meet the time level t=j must be contained in the interval between the two end points of the numerical algorithm.

In other words: *the domain of dependence of the differential equation should be contained within the domain of dependence of the discretized equation.*

This profound observation was first made in a pioneering research paper by Richard Courant, Kurt Otto Friedrichs and Hans Lewy[50] entitled 'Uber die partiellen differenz-gleichungen der mathematischen Physik' which was published in the journal *Mathematische Annalen* in 1928[51]. As the title suggests, the paper covers a wide spectrum of problems relating to the wave equation, the Laplace equation, etc. - and in particular considers the question of 'consistency' of a numerical scheme, that is, its ability to approach the exact solution of the differential equation as the mesh size tends to zero. In the case of hyperbolic equations, the authors showed that a numerical scheme could not be consistent if the numerical domain of dependence did not contain the physical domain of dependence. Indeed, without actually having to refer to their paper, the reader can visualize the logic: suppose the initial data which defines the uniqueness is altered in the region outside the numerical - but within the physical - domain of dependence. Then the true solution will change but the numerical solution will not reflect these changes - and no amount of grid refinement will help.

In acknowledgement of the pioneering work of Courant, Friedrichs and Lewy, the requirement that the domain of dependence of the differential equation be contained in the domain of dependence of the discretized equation is called the CFL condition. While this condition does not guarantee stability - except in the case of

---

[49] In general these characteristics will not be straight lines, but they may be approximated by straight lines from one time-step to the next.

[50] See footnote 45 for details about Courant. Friedrichs and Lewy were his students, both of whom completed their Ph.D degrees at the University of Gottingen in 1925/26. A decade later, all three emigrated to the USA. Friedrichs helped Courant found the Institute for Mathematics and Mechanics in New York and in 1948, they wrote the famous book *Supersonic Flow and Shock Waves*. Lewy joined the University of California at Berkeley.

[51] An English translation, 'On the partial differential equations of mathematical physics' appeared in the *IBM Journal* in 1967.

simple schemes such as (3.43) and (3.45) - we may be certain that any scheme that does not satisfy this criterion will be unstable.

Returning now to the first order system (1.21), we see that for the CFL condition to be satisfied, the time step must satisfy the inequality $\Delta t < \Delta x / |\lambda_{max}|$, where λ_{max} is the eigenvalue whose absolute value is largest. The modulus of this eigenvalue is known as the spectral radius of the matrix $[\mathbf{F}']$ and is a very important concept in the stability theory of general systems of equations. Readers wishing to pursue this subject could refer to the text book by Richtmyer and Morton (1967), a classic work, to which we have already had occasion to refer to in the first chapter.

One further point deserves comment. The reader will surely be aware that implicit schemes are always stable regardless of the value of the Courant number. Obviously they must satisfy the CFL condition. To see how this is so, we first note that any implicit scheme has at least two grid points at the time-level j+1. Because the solution at both these points are unknown, we have no choice but to solve the whole set of discretized equations simultaneously. In other words, the solution at *any* point on the time level j+1 will depend on the solution at *every* point at time-level j, so that the numerical domain of dependence (which is in fact the entire domain) invariably contains the physical domain of dependence.

Thus far we have only been considering problems in one space dimension. What form does the CFL condition take in two space dimensions? Consider the two-dimensional form of the linear advection equation

$$u_t + \lambda_1 u_x + \lambda_2 u_y = 0 \qquad \ldots (3.47a)$$
$$u(x,y,0) = f(x,y). \qquad \ldots (3.47b)$$

The solution to this initial value problem is the two-dimensional counterpart of (3.37), that is

$$u(x,y,t) = f(x-\lambda_1 t, y-\lambda_2 t). \qquad \ldots (3.48)$$

In other words the solution is propagated by straight lines in 3-D space having the directional derivatives $dx/dt = \lambda_1$ and $dy/dt = \lambda_2$. The general form of the explicit discretization scheme - analogous to (3.38) - will be

$$u(i,j,k+1) = \Sigma c_{mn} u(i+m,j+n,k), \qquad \ldots (3.49)$$

where the coefficients c_{mn} are functions of the two Courant numbers $\sigma_1 = \lambda_1 \Delta t/\Delta x$ and $\sigma_2 = \lambda_2 \Delta t/\Delta y$. The special case of first order backward differencing - analogous to (3.43) - will be

$$u(i,j,k+1) = \sigma_1 u(i-1,j,k) + \sigma_2 u(i,j-1,k) + (1-\sigma_1-\sigma_2)u(i,j,k). \qquad \ldots (3.50)$$

Just as with (3.43), this algorithm should not be employed unless both λ_1 and λ_2 are positive. Otherwise forward differencing should be employed in the relevant direction (or in both if necessary).

The amplification factor for the general formula (3.49) - analogous to (3.42) - is:

$$AF(\theta_1,\theta_2) = \Sigma c_{mn}\exp\{i(m\theta_1+n\theta_2)\} \qquad \ldots (3.51)$$

which in the particular case of (3.50) reduces to

$$AF(\theta_1,\theta_2) = \sigma_1\exp(-i\theta_1) + \sigma_2\exp(-i\theta_2) + (1-\sigma_1-\sigma_2). \qquad \ldots (3.52)$$

analogous to (3.44). We now have two frequencies θ_1 and θ_2 and we require the amplification factor to be less than one, whatever be the value of either frequency. Separating (3.52) into real and imaginary parts ξ and η, we get

$$\xi = 1 - \sigma_1 - \sigma_2 + \sigma_1\cos\theta_1 + \sigma_2\cos\theta_2 \qquad \ldots (3.53a)$$
$$\eta = -\sigma_1\sin\theta_1 - \sigma_2\sin\theta_2. \qquad \ldots (3.53b)$$

This is a two-parameter family. Without loss of generality, we may take σ_1 as the first parameter: it describes a circle in the complex plane centered at $(1-\sigma_1-\sigma_2,0)$ with radius σ_1. The second parameter describes another circle centred at any point on the first circle. For the amplification factor to be less than one, the second parameter must always lie within the unit circle. It is not hard to see that the condition for this is

$$\sigma_1 + \sigma_2 < 1. \qquad \ldots (3.54)$$

Thus the condition is rather more stringent than one might have expected; it is not enough for the two Courant numbers to be less than unity; so must their sum. Naturally this leads to a smaller allowable time-step; in fact from (3.54):

$$\lambda_1(\Delta t/\Delta x) + \lambda_2(\Delta t/\Delta y) < 1 \qquad \ldots (3.55)$$

yielding

$$\Delta t < \frac{\Delta x\Delta y}{\lambda_1\Delta y + \lambda_2\Delta x} \qquad \ldots (3.56)$$

and the right-hand-side is clearly smaller than both $(\Delta x/\lambda_1)$ and $(\Delta y/\lambda_2)$.

How does inequality (3.56) relate to the CFL condition which requires the physical domain of dependence to be contained in the numerical domain of dependence? Consider Figure 3.7b (page 63). The solution is being computed at the point $(i,j,k+1)$ - which is above the plane of the paper - and the three points on the plane of the paper which contribute to this solution are $(i-1,j,k)$, $(i,j-1,k)$ and (i,j,k) (denoted, in Figure 3.7b, by the letters X, Y and O). The numerical domain of dependence is simply the triangle formed by these three points. To use our previous analogy, it is the 'firm ground' on which the stalk that supports the solution at $(i,j,k+1)$ should rest. This stalk is, of course, simply the characteristic passing through the point $(i,j,k+1)$; it has the directional derivatives $dx/dt = \lambda_1$ and $dy/dt =$

65

λ_2. Thus if we assume, for the sake of convenience, that the Cartesian coordinates at the point (i,j,k) are (0,0), then the characteristic passing through (i,j,k+1) meets the plane of the paper at the point A with coordinates $(-\lambda_1\Delta t,-\lambda_2\Delta t)$ (Figure 3.7b).

Let us draw a straight line from the origin through the point $(-\lambda_1\Delta t,-\lambda_2\Delta t)$ and let it meet XY at B. Construct BC parallel to XO. In order that the point A lie within the triangle XOY, we should have $\lambda_1\Delta t <$ BC and $\lambda_2 \Delta t <$ OC. Now because triangles XOY and BCY are similar, we have

$$BC/XO = CY/OY \text{ or } BC/\Delta x = (\Delta y - OC)/\Delta y$$
$$\text{or } BC\Delta y + OC\Delta x = \Delta x\Delta y. \qquad \dots (3.57)$$

Again, because the points O, A and B are on one straight line:

$$BC/OC = (\lambda_1\Delta t)/(\lambda_2\Delta t) = \lambda_1/\lambda_2 \qquad \dots (3.58)$$

and substitution into (3.57) yields

$$BC = \frac{\lambda_1\Delta x\Delta y}{\lambda_1\Delta y + \lambda_2\Delta x} \quad : \quad OC = \frac{\lambda_2\Delta x\Delta y}{\lambda_1\Delta y + \lambda_2\Delta x} \qquad \dots (3.59)$$

The conditions $\lambda_1\Delta t <$ BC and $\lambda_2\Delta t <$ OC both reduce to inequality (3.56).

Thus, as in the one-dimensional case, the CFL condition requires precisely the same restriction as the Von Neumann stability criterion. But once again it must be stressed that this equivalence holds only for simple algorithms such as (3.50).

We are now ready to move on to the final stage in the heirarchy - as far as the present work is concerned - that of systems of equations in two dimensions. Here again we have not one characteristic, but several. The theory of characteristics for hyperbolic systems in two independent variables is far more complicated than its counterpart in one dimension; interested readers could refer to the relevant chapters in the books by Garabedian (1964) and Richtmyer & Morton (1968)[52]. Here let us just say that the set of characteristics passing through the point (i,j,k+1) form a conoid which cuts the time-level 'k' in a closed curve. The CFL condition requires that this entire closed curve be contained within the numerical domain of dependence, that is to say the polygon formed by joining all the extreme grid points in the numerical algorithm.

In order to satisfy the CFL condition for two-dimensional systems, two precautions must be taken. In the first place the numerical algorithm must contain at least as many grid points as are necessary to ensure that, for a sufficiently small time-step the closed curve can be contained in the polygon generated by these grid points. Thus the backward-differencing algorithm (3.50) will not suffice for a closed curve which encloses the point (i,j,k) regardless of how small a time step is

---

[52] Garabedian (Chapter 6); Richtmyer and Morton (Chapter 13).

chosen. However the reader can easily see that a five-point algorithm making use of the points (i-1,j,k), (1,j-1,k), (i,j,k), (i,j+1,k), (i+1,j,k) will always suffice provided the time-step is sufficiently small. This brings us to our second precaution. Just how small should the time-step be? On the basis of our analysis in connection with the simple two-dimensional advection equation (3.47), we see that if any particular line of our characteristic conoid has the directional derivatives dx/dt = λ_1 and dy/dt = λ_2, then if λ_1 and λ_2 are both positive, the condition for that line to fall within the numerical domain of dependence would be given by the inequality (3.56). If either, or both, of λ_1 and λ_2 is negative, they must be replaced by their absolute values in this inequality. Therefore if λ_{1max} and λ_{2max} denote the maximum values of these directional derivatives, the general condition for the characteristic conoid to fall within the numerical domain of dependence will be:

$$\Delta t < \frac{\Delta x \Delta y}{|\lambda_{1max}|\Delta y + |\lambda_{2max}|\Delta x} \qquad \ldots (3.60)$$

In the case of the Euler equations, the four eigenvalues of the matrix $\underline{F}'$ (see page 51) are given by u, u, u+c and u-c (this was shown in Topic 3: Eq 3.24). As both u & c are positive, the largest of these eigenvalues is u+c. Similarly the largest of the four eigenvalues of the matrix $\underline{G}'$ is v+c. Thus the CFL criterion for a numerical on the Euler equations takes the form:

$$\Delta t < \frac{\Delta x \Delta y}{(u+c)\Delta y + (v+c)\Delta x} \qquad \ldots (3.61)$$

In the actual Fortran program, the stability criterion (3.61) will have to be modified to account for the fact that the coordinate system is not Cartesian but body-fitted. This is not very difficult and we will come to it in Topic 10 on general programming details. But for the moment, we will leave the subject of stability and move on to the next important topic: space-discretization.

Topic 6. Space-discretization: updating the fundamental variables

Having set the value of the time-step, we now have all the data we require to commence work on the 'central core' of the solution procedure: the updating of the fundamental variables ρ, ρu, ρv and ρe. Each of these fundamental variables is stored in the centre of a particular cell or 'control volume' and they will be updated by means of the so-called finite volume formulation of the governing equations. The philosophy behind this formulation is well known; the method has been around for at least two decades[53] and is now considered just as basic as the classical finite difference approach. Indeed in many cases, the two formulations may actually be shown to yield equivalent results - in much the same way as the iterative and artificial time-marching procedures were shown to be equivalent for the boundary value problem (1.6) (see page 6).

In order to impose the finite volume approach on a system such as (3.15) (page 49), it is first necessary to express it in an integral formulation. Let us integrate it over a typical control volume, that is to say the cell in our grid at which the fundamental variables are to be updated. Denoting its volume[54] as V, Equation (3.15) can be expressed as:

$$\partial/\partial t \ [\textstyle\int U dV] \ + \ \textstyle\int [F_e - F_w] dY \ + \ \textstyle\int [G_n - G_s] dX \ = \ 0, \qquad \ldots (3.62)$$

where the suffixes e & w denote the easternmost and westernmost values of X for a particular value of Y in the cell under consideration; similarly n & s denote the northernmost and southernmost values of X for a particular Y. If the grid is Cartesian, the X-coordinates of e & w, and the Y-coordinates of n & s, will be fixed for each cell, but in general they will vary.

The underlying assumption in the finite-volume approach is that each of the variables may be approximated by a constant in a particular control volume, this constant being set equal to the value of the variable at the centre of the cell. Thus the integro-differential equation, (3.62), effectively reduces to a straightforward process of subtraction, multiplication and addition. To illustrate this, let us first assume that the grid is Cartesian so that each control volume (or grid-cell) is simply a rectangle with sides Δx and Δy (and volume ΔV). Then with the time-step Δt computed by the method of the previous section, we have

---

[53] The two papers generally credited with the introduction of the finite volume approach in CFD are McDonald (1971), and MacCormack and Paullay (1972), both of whose authors worked independently. Rizzi and Inouye (1973) extended the technique to three dimensions. In honour of his pioneering work, MacCormack was invited to present the keynote lecture at the 1993 AIAA conference on CFD held at Orlando, Florida: see MacCormack (1993).

[54] Strictly we ought to say 'area', this being two dimensions, but 'volume' has now become the jargon word for both two and three dimensions, in keeping with the descriptive terms finite-volume and control-volume. See also footnote 32, page 39.

$$(\Delta V/\Delta t)[U^{t+1}-U^t] = -\{(\Delta y)[F_e-F_w]^t+(\Delta x)[G_n-G_s]^t\} \qquad \cdots \quad (3.63)$$

which can be expressed in either of the two forms:

$$U^{t+1} = U^t - \{(\Delta y)[F_e-F_w]^t+(\Delta x)[G_n-G_s]^t\}(\Delta t/\Delta V) \qquad \cdots \quad (3.64a)$$

or

$$U^{t+1} = U^t + \{(\Delta y)[F_w-F_e]^t+(\Delta x)[G_s-G_n]^t\}(\Delta t/\Delta V) \qquad \cdots \quad (3.64b)$$

Here t denotes the time level at which the flow is known; (t+1) the next time-level where the flow is to be computed. Numerically it makes no difference which formula is used, but (3.64b) is slightly more meaningful as the second term on its right hand side represents the *increment* by which U is to be updated at each time-step. Thus in the case of a Cartesian grid, the four fundamental variables may be updated by the straightforward application of Equation (3.64b), by substituting the appropriate components of the vectors U, F and G for the Euler equations, which are given by (3.16) (page 49). More specifically:

$$\rho^{t+1} = \rho^t + \{(\Delta y)[(\rho u)_w-(\rho u)_e]^t+(\Delta x)[(\rho v)_s-(\rho v)_n]^t\}(\Delta t/\Delta V) \qquad (3.65a)$$

$$(\rho u)^{t+1} = (\rho u)^t + \{(\Delta y)[(\rho u^2+p)_w-(\rho u^2+p)_e]^t+(\Delta x)[(\rho vu)_s-(\rho vu)_n]^t\}(\Delta t/\Delta V) \qquad (3.65b)$$

$$(\rho v)^{t+1} = (\rho v)^t + \{(\Delta y)[(\rho uv)_w-(\rho uv)_e]^t+(\Delta x)[(\rho v^2+p)_s-(\rho v^2+p)_n]^t\}(\Delta t/\Delta V) \qquad (3.65c)$$

$$(\rho e)^{t+1} = (\rho e)^t + \{(\Delta y)[(\rho uh)_w-(\rho uh)_e]^t+(\Delta x)[(\rho vh)_s-(\rho vh)_n]^t\}(\Delta t/\Delta V). \qquad (3.65d)$$

Two comments should be made here. One is that the variables involved are stored only at the cell centres and therefore will first have to be computed at the cell faces (by straightforward interpolation between neighbouring cells) before Eqs (3.65) can be used for the purpose of updating. Second, because all four of these equations involve the products $(\rho u)(\Delta y)$ and $(\rho v)(\Delta x)$, it is convenient (and time-saving) to store these quantities before the updating process begins. Physically, these products are the mass fluxes in the X and Y directions - and if we denote them by Q_1 & Q_2 respectively, Equations (3.65) take the form:

$$\rho^{t+1} = \rho^t + \{[(Q_1)_w-(Q_1)_e]^t+[(Q_2)_s-(Q_2)_n]^t\}(\Delta t/\Delta V) \qquad \cdots \quad (3.66a)$$

$$(\rho u)^{t+1} = (\rho u)^t + \{[(uQ_1+p)_w-(uQ_1+p)_e]^t+[(uQ_2)_s-(uQ_2)_n]^t\}(\Delta t/\Delta V) \qquad \cdots \quad (3.66b)$$

$$(\rho v)^{t+1} = (\rho v)^t + \{[(vQ_1)_w-(vQ_1)_e]^t+[(vQ_2+p)_s-(vQ_2+p)_n]^t\}(\Delta t/\Delta V) \qquad \cdots \quad (3.66c)$$

$$(\rho e)^{t+1} = (\rho e)^t + \{[(hQ_1)_w-(hQ_1)_e]^t+[(hQ_2)_s-(hQ_2)_n]^t\}(\Delta t/\Delta V). \qquad \cdots \quad (3.66d)$$

It is worth pausing at this point to understand the physical significance of these equations. Let us start with Equation (3.66a). The quantities $(Q_1)_w$ and $(Q_1)_e$ denote the mass flow entering the west face of the cell and leaving the east face. It is of course quite possible for the flow to enter through the east face and leave through the west face; this will happen in a zone of flow-reversal. In such a case, $(Q_1)_w$ and $(Q_1)_e$ will both be negative, but the validity of the equations will not be affected. Similarly the quantities $(Q_2)_s$ and $(Q_2)_n$ denote the mass fluxes entering the south face and leaving the north face; they may be either positive or negative depending upon whether the flow is inclining upwards or downwards at the cell under consideration. For a typical turbine problem (Figure 3.8a, page 71) the streamlines incline from north-west to south-east so that (assuming that the grid is Cartesian[55]) the flow would tend to enter from the north and west faces and leave through the south and east faces. For a typical compressor problem (Figure 3.8b), the streamlines are directed from south-west to north-east so that the flow would tend to enter from the south and west faces and leave through the north and east faces.

If we discount, for the moment, the volume ΔV, then the right hand side of Equation (3.66a) is simply the net increase in mass in the cell over the time-interval Δt. Dividing this by ΔV yields the net increase in density. In other words Equation (3.66a) is simply an algebraic expression of the law of conservation of mass in the 'space-time capsule' with dimensions Δx, Δy and Δt. Obviously this must be so, as it was derived from the conservation-of-mass component of the Euler equations, but it is interesting to note (as we have just shown) that it can also be derived by pure reasoning. Similarly Equations (3.66b) & (3.66c) express conservation of momentum and Equation (3.66d) expresses conservation of energy.

Equations (3.66) have been derived under the assumption that the grid is Cartesian but of course this will not generally be the case. So let us now see how the equations complexify, assuming the cell under consideration is an arbitrary quadrilateral.

Let us return to Figure 3.5a (page 41). Here we do not have any specific north, south, east and west faces, but Equation (3.62) (page 68) is still valid except that the suffixes w & e now refer to the westernmost and easternmost points for a given Y-coordinate and s & n refer to the southernmost and northernmost points for a given X-coordinate. Thus in order to evaluate $\int F_w dy$, we will have to integrate along both CA & AB; to evaluate $\int F_e dy$, we will have to integrate along CD & DB; to evaluate $\int G_s dx$, we will have to integrate along AC & CD; and to evaluate $\int G_n dx$, we will have to integrate along AB & BD. If we denote by F_{AC} and G_{AC} the values of F and G along the face AC (and similarly for the other faces), then:

$$\int F_w dy = F_{AC}(GA) + F_{AB}(AE) = F_{AC}[Y(A)-Y(C)] + F_{AB}[Y(B)-Y(A)] \quad \text{... (3.67a)}$$
$$\int F_e dy = F_{CD}(HD) + F_{BD}(DF) = F_{CD}[Y(D)-Y(C)] + F_{BD}[Y(B)-Y(D)] \quad \text{... (3.67b)}$$
$$\int G_s dx = G_{AC}(GC) + G_{CD}(CH) = G_{AC}[X(C)-X(A)] + G_{CD}[X(D)-X(C)] \quad \text{... (3.67c)}$$
$$\int G_n dx = G_{AB}(EB) + G_{BD}(BF) = G_{AB}[X(B)-X(A)] + G_{BD}[X(D)-X(B)]. \quad \text{... (3.67d)}$$

---

[55] Of course if the grid is body-fitted, the streamwise grid–lines would automatically align with the flow direction; in this case the Q_2 mass fluxes will be small compared to the Q_1 mass fluxes and may be either positive or negative.

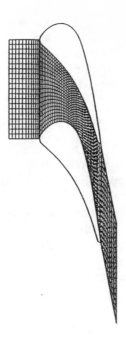

Figure 3.8a   **A typical turbine problem**
(streamlines directed from northwest to southeast)

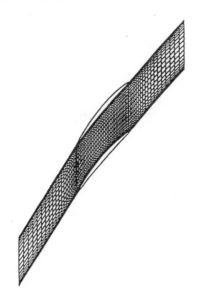

Figure 3.8b   **A typical compressor problem**
(streamlines directed from southwest to northeast)

The factors GA, AE, HD, DF, GC, CH, EB & BF are the projective lengths which formed part of the discussion in Topic 2. All are stored at the start of the program, though (as explained on page 39) only four arrays are actually needed.

Substitution of (3.67) in (3.62) yields the general formula for updating the vector **U** - analogous to Equation (3.64b):

$$\mathbf{U}^{t+1} = \mathbf{U}^t + [\mathbf{F}_{in} - \mathbf{F}_{out} + \mathbf{G}_{in} - \mathbf{G}_{out}]^t \, (\Delta t/\Delta V) \qquad \ldots \ (3.68)$$

where $\mathbf{F}_{in}$, $\mathbf{F}_{out}$, $\mathbf{G}_{in}$ and $\mathbf{G}_{out}$ are given by the right hand sides of Eqs (3.67a), (3.67b), (3.67c) and (3.67d) respectively. The vectors $\mathbf{F}_{AC}$, $\mathbf{F}_{AB}$, $\mathbf{F}_{CD}$, $\mathbf{F}_{BD}$, $\mathbf{G}_{AC}$, $\mathbf{G}_{CD}$, $\mathbf{G}_{AB}$ & $\mathbf{G}_{BD}$, which appear in these equations, can be obtained by interpolation from the stored values of **F** & **G** at the cell-centres at time-level t.

In the actual program, four subroutines will be employed corresponding to the four fundamental variables ρ, ρu, ρv and ρe. The purpose of each of these subroutines is to determine the increments $\Delta\rho$, $\Delta\rho u$, $\Delta\rho v$ and $\Delta\rho e$ - given by the second term on the right hand side of Equation (3.68) - by which these fundamental variables are to be updated. To update ρ, we substitute $\mathbf{F}_1 = Q_1$ and $\mathbf{G}_1 = Q_2$ to obtain:

$$\Delta\rho = \{[(Q_1)_{in}-(Q_1)_{out}] + [(Q_2)_{in}-(Q_2)_{out}]\} \, (\Delta t/\Delta V). \qquad \ldots \ (3.69a)$$

To update ρu, we substitute $\mathbf{F}_2 = uQ_1+p$ and $\mathbf{G}_2 = uQ_2$ to obtain:

$$\Delta\rho u = \{[(uQ_1+p)_{in}-(uQ_1+p)_{out}] + [(uQ_2)_{in}-(uQ_2)_{out}]\} \, (\Delta t/\Delta V). \qquad \ldots \ (3.69b)$$

To update ρv, we substitute $\mathbf{F}_3 = vQ_1$ and $\mathbf{G}_3 = vQ_2+p$ to obtain:

$$\Delta\rho v = \{[(vQ_1)_{in}-(vQ_1)_{out}] + [(vQ_2+p)_{in}-(vQ_2+p)_{out}]\} \, (\Delta t/\Delta V). \qquad \ldots \ (3.69c)$$

To update ρe, we substitute $\mathbf{F}_4 = hQ_1$ and $\mathbf{G}_4 = hQ_2$ to obtain:

$$\Delta\rho e = \{[(hQ_1)_{in}-(hQ_1)_{out}] + [(hQ_2)_{in}-(hQ_2)_{out}]\} \, (\Delta t/\Delta V). \qquad \ldots \ (3.69d)$$

Having derived these four key equations, we now move on to the next topic.

Topic 7. Artificial dissipation

Equations (3.69) effectively represent the 'building blocks' of our solution procedure - or, if we return to our analogy with the construction of a building, these four equations may be thought of as four pillars which rest on one floor (that is, time level) and support the next. From a purely theoretical view point, these four equations are all that are required for the formulation of a solution algorithm. Starting with an initial guess at time-level t = 0, and selecting a time step in accordance with the CFL criterion, we update the four fundamental variables ρ, ρu, ρv & ρe using Equations (3.69 a,b,c,d) and so compute all the relevant variables at the first time-level. The updating can be done in a single stage (as in a classical numerical solution to an ODE: see page 1) or in several stages (as in a Runge-Kutta solution technique). After application of appropriate boundary conditions, we can repeat the procedure from time-steps 1 to 2, then from time-steps 2 to 3, and so on until (hopefully) the process converges.

This is the theory. Let us see what happens in practice. Figures 3.9 (page 75) depict the Mach number distribution at different time-levels if the algorithm just described is applied to a circular arc cascade. Small oscillations appear in the solution at every time step which, instead of dying away, become worse as the algorithm progresses through increasing levels of time. In effect the solution procedure is showing classical symptoms of instability. But why should this be the case when the time-steps have been chosen in accordance with the CFL condition and the Von Neumann stability criterion?

The answer to this question lies in the fact that these 'classical' criteria for stability are strictly valid only for linear equations. If the equations are nonlinear, then the analysis becomes far more complicated and very difficult to pursue rigorously. In fact the field of nonlinear mathematics is an area where numerical experience is of more practical significance than theoretical considerations. This statement is not made with a view to undermining the importance of theoretical analysis; conditions such as the Von Neumann and CFL criteria must certainly hold if the solution procedure is to have any chance of converging. But they may not be able to stand on their own - they may have to be backed up by other stabilizing factors. Continuing our analogy with the construction of a building, the linearity or nonlinearity of the equations may be compared to the regularity or irregularity of the earth below the building. If the earth is regular, the building should be stable provided the distance between the floors is not too large (the Von Neumann or CFL condition). But if the earth is irregular (perhaps earthquake-prone), the structure may prove unstable no matter how closely the floors are built. This is not to say that buildings cannot be constructed on earthquake-prone ground; it simply means that the pillars which support each floor must be strengthened by the application of earthquake-proof material. But how does one obtain the earthquake-proof material? This is where experience is of great importance. And in the case of a time-marching algorithm, experience has shown that the factor that stabilizes the numerical scheme is a mathematical expression that is universally known by the epithets 'artificial dissipation' or 'artificial viscosity'.

The concept of artificial viscosity was introduced - appropriately enough - by Von Neumann in a classic paper[56] with R.D.Richtmyer (1950). It is most easily understood in the context of two basic equations that were discussed in the introductory chapter of this work: the nonlinear advection equation (1.16) and Burger's equation (1.19) which, for the sake of convenience, are repeated here:

Nonlinear advection equation	$u_t + uu_x = 0$	. . . (3.70)
Burger's equation	$u_t + uu_x = vu_{xx}.$	. . . (3.71)

These two equations are fundamentally very different. At first sight it may appear that (3.71) would present more difficulties in the solution procedure as it contains a diffusion term in addition to the time-dependent term and convection term that are common features of both equations. Superficially, the two equations may be likened to embryonic forms of the Euler and Navier-Stokes equations respectively - and we all know that the Navier-Stokes equations are the much more complex of the two.

Nevertheless, strange as it may seem, numerical solutions of (3.71) are actually easier to generate than those of (3.70). The diffusion term has a stabilizing influence - so much so that a numerical solution algorithm for (3.70) can actually be strengthened by the addition of an artificial diffusion term - which ideally should be just large enough to stabilize the solution. In practice, therefore, numerical solution techniques for Equations (3.70) and (3.71) run along parallel lines despite the fundamental differences in the nature of the equations.

Does this mean that solution techniques for the Euler and Navier-Stokes equations must also follow parallel lines? Well not exactly. It is true that some kind of artificial viscosity will have to be imposed on the numerical solution procedure for the Euler equations, but this need be nothing quite so complex as the genuine viscosity terms which typify the Navier-Stokes equations even in the relatively simple case of laminar flow (and if the flow is turbulent, a turbulence model would have to be set up; this may mean the implementation of additional equations into the system). It is important to remember that the artificial viscosity that is associated with an Euler solver plays an entirely different role to that of real viscosity; it is present purely and simply for the purpose of stabilizing the algorithm. Thus there is absolutely no need to compute such terms accurately[57] - as one would naturally try to do in the case of the genuinely viscous terms in the Navier-Stokes equations.

---

[56] The reader may recall that no classic paper exists with regard to Von Neumann's stability criterion: footnote 46, page 60.

[57] Indeed the concept of accuracy is somewhat meaningless here as there is no term in the governing equations that the artificial viscosity is attempting to model.

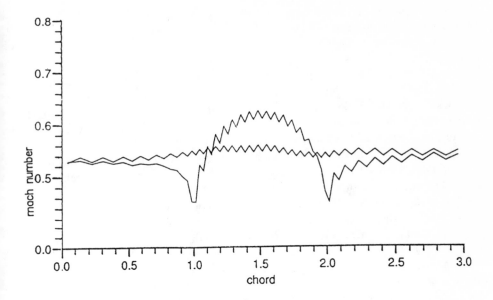

Figure 3.9a A typical Mach number distribution which may arise after about 1000 time-steps if no artificial viscosity is employed

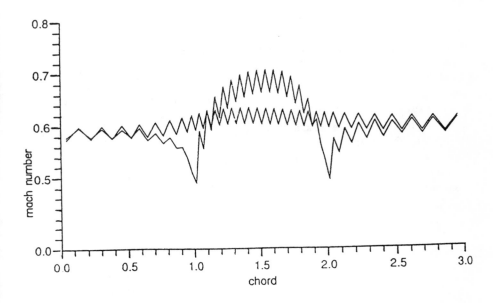

Figure 3.9b A typical Mach number distribution which may arise after about twice as many time-steps as in Figure 3.9a

Ideally the artificial viscosity terms should be as simple as possible - provided they serve their purpose. Of course, this is easier said than done and, as a result, a great deal of research has been devoted to numerical experiments with different kinds of artificial viscosity terms. It is interesting to note that the genuinely viscous terms of the Navier-Stokes equations may not in fact have adequate stabilizing effects (despite their complexity) and, especially in the presence of shocks, a Navier-Stokes solver will, in general, need to incorporate artificial viscosity in addition to the 'real thing'. For subcritical flows however, the stabilizing effect of the genuinely viscous terms is usually adequate.

Since the appearance of the original paper by Von Neumann, the concept of artificial viscosity has been discussed and/or utilized by several researchers, of which perhaps the most important is that of Lax and Wendroff (1960), who provide a detailed theoretical discussion of the corrective role played by artificial viscosity[58]. In particular they show how certain schemes exhibit error modes which change sign at alternate grid points leading to the oscillatory type of numerical solution seen in Figures.3.9 and how the artificial viscosity tends to counteract these oscillations by simulating the effects of physical viscosity *on the scale of the mesh.*

Later work on artificial viscosity, especially since the high-speed computer era, has tended to concentrate more on the practical than theoretical: the search for a dissipation operator that would do its job effectively in the context of a given problem. Note that the application of artificial dissipation is by no means restricted to problems involving time-marching. In the 1970s when transonic potential solvers more-or-less represented the frontiers of CFD research, artificial viscosity techniques were often employed to aid shock-capturing. Indeed even the classic paper of Murman and Cole (1971)[59] made use of an artificial viscosity of sorts - that is to say their upwind differencing (which they applied in the supersonic zone) was equivalent to a central differencing (which they applied in the subsonic zone) plus a dissipation operator. Murman and Cole solved the small disturbance equation, which is an approximate form of the full potential equation, but later in the decade, similar techniques were employed to solve the full potential equation as well. In the context of the Euler / Navier-Stokes equations, artificial dissipation was probably first used by MacCormack and Baldwin (1975). In all these instances, there has been no 'law' by which the dissipation operator was derived; it was simply a matter of trial and error - and making use of the operator which worked most effectively.

---

[58] The reader may recall that the work of Lax and his colleagues lays stress on such theoretical discussions: see page 11.

[59] Normally when we speak of classic papers, we are inclined to think of papers at least half a century old. But the paper of Murman and Cole, despite being relatively recent, represented an important breakthrough; their idea of combining different methods of differencing resulted in the first successful computation of a steady transonic flow for the small perturbation equation.

Let us summarise what we have learnt about artificial viscosity. First it is the additional stabilizing factor that is required for nonlinear problems[60] (apart from the CFL and Von Neumann conditions which must hold regardless of linearity). There is no hard-and-fast rule about the form it should take; it is simply a matter of trying out different operators and seeing how well they perform for a given problem. This is an appropriate moment to quote Jameson (1982) - from a lecture that he delivered at a conference: 'The dissipation operator which I have found to work is actually a combination of second and fourth differences'. The problem that Jameson was attempting to solve was very similar to the current work: the solution of the unsteady Euler equations using time-marching.

The dissipation operator used by Cheng (and described here) is very similar to that of Jameson. And if the reader is inclined to criticise Cheng for simply adapting the operator of Jameson rather than finding one for himself, he should be reminded that research is a cooperative effort and that if progress is to be made in any field, one must build upon the work of others. And unlike so many other present-day researchers who have the good fortune to possess a code right from the start which they can then proceed to expand or refine, Cheng had to create his code right from scratch. When one takes into account the fact that in the three years in which he worked towards a Ph.D degree, he was able to develop a code that could deal with the three-dimensional Navier-Stokes equations with up-to-date turbulence models, it is hardly surprising that he had to base some of his algorithms on work which had already been done by others.

In the present program, the dissipation operations have been succinctly expressed in the form of two subroutines named ARTCF (artificial coefficients) and ARTVIS (artificial viscosity). The first evaluates four arrays which store the coefficients of the second and fourth order dissipation operators in the two grid-coordinate directions. The second computes the dissipation terms themselves; these also require four arrays, one corresponding to each of the four fundamental variables. If we denote by $A\rho$, $A\rho u$, $A\rho v$ & $A\rho e$ the dissipative contribution, then these must be added to $\Delta\rho$, $\Delta\rho u$, $\Delta\rho v$ & $\Delta\rho e$ (see Equations 3.69) and the respective sums will form the basis for the updating of the four fundamental variables.

Let us now provide the specific formula by which $A\rho$, $A\rho u$, $A\rho v$ & $A\rho e$ are computed. In fact if we denote by f any of the four fundamental variables, then

$$Af = a_{x1}{}^+f_{x1}{}^+ - a_{x1}{}^-f_{x1}{}^- - a_{x3}{}^+f_{x3}{}^+ - a_{x3}{}^-f_{x3}{}^- + a_{y1}{}^+f_{y1}{}^+ - a_{y1}{}^-f_{y1}{}^- - a_{y3}{}^+f_{y3}{}^+ - a_{y3}{}^-f_{y3}{}^- \quad (3.72)$$

where the superscripts + and - indicate that the concerned terms have been

---

[60] Even in the case of linear problems, artificial dissipation serves the useful purpose of damping the oscillations which appear in the vicinity of discontinuities in central difference schemes. Of course, discontinuities cannot arise in linear problems unless they are present in the initial data. The so-called Riemann problem, which is the linear advection equation with initial data given by two constants meeting at a 'shock' forms a very basic test case on which to test different methods of artificial dissipation.

computed at the positive and negative walls of the cell (i.e +ve east, -ve west in the x direction, +ve north, -ve south in the y direction). The terms involving the suffix x are specifically defined below; those involving the suffix y may be defined analogously.

$$a_{x1}^{+} = a_1 max[p_{xx}(i,j),p_{xx}(i+1,j)] \qquad \dots (3.73a)$$
$$a_{x1}^{-} = a_1 max[p_{xx}(i,j),p_{xx}(i-1,j)] \qquad \dots (3.73b)$$

where

$$p_{xx}(i,j) = \frac{p(i+1,j)-2p(i,j)+p(i-1,j)}{p(i+1,j)+2p(i,j)+p(i-1,j)} \qquad \dots (3.74)$$

$$a_{x3}^{+} = max[(a_3-a_{x1}^{+}),0] \quad ; \quad a_{x3}^{-} = max[(a_3-a_{x1}^{-}),0] \qquad \dots (3.75)$$

$$f_{x1}^{+} = f(i+1,j) - f(i,j) \quad ; \quad f_{x1}^{-} = f(i,j) - f(i-1,j) \qquad \dots (3.76)$$

$$f_{x3}^{+} = f(i+2,j) - 3f(i+1,j) + 3f(i,j) - f(i-1,j) \qquad \dots (3.77a)$$
$$f_{x3}^{-} = f(i+1,j) - 3f(i,j) + 3f(i-1,j) + f(i-2,j). \qquad \dots (3.77b)$$

Note that a_{x3}^{+} and a_{x3}^{-} are defined in such a manner so as to get switched off in regions where a_{x1}^{+} and a_{x1}^{-} are large, i.e near shocks - as they are found to cause spurious oscillations. The coefficients a_1 and a_3 must be chosen by the programmer.

Topic 8. Time-discretization: the Runge-Kutta integration process

Let us go back, momentarily, to the very first page of this book. We started out with the ODE (1.1) and its classic method of solution (1.3). The values of 'y' are being updated in steps of 'h' by the quantity h{f(x,y)}.

For our present problem, the quantities $\Delta\rho$, $\Delta\rho u$, $\Delta\rho v$ & $\Delta\rho e$ given by Equations (3.69a,b,c,d) (page 72) are analogous to the quantity h{f(x,y)} of Equation (1.3) (page 1). Well not quite. As we have explained in the previous section, they must also be augmented by the artificial dissipation terms $A\rho$, $A\rho u$, $A\rho v$ & $A\rho e$ defined by equations (3.72) to (3.77). But having done so, the four fundamental variables ρ, ρu, ρv & ρe - which are analogous to 'y' in Equation (1.1) - may be updated analogously to (1.3), that is:

$$\rho(t+\Delta t) = \rho(t) + \Delta\rho + A\rho \qquad \dots (3.78a)$$
$$\rho u(t+\Delta t) = \rho(t) + \Delta\rho u + A\rho u \qquad \dots (3.78b)$$
$$\rho v(t+\Delta t) = \rho(t) + \Delta\rho v + A\rho v \qquad \dots (3.78c)$$
$$\rho e(t+\Delta t) = \rho(t) + \Delta\rho e + A\rho e \qquad \dots (3.78d)$$

Now let us return, momentarily, to page 2 of this book. For convenience, we repeat a section of the second paragraph:

Note, incidentally, that the classical method of solution shown above is only of first order accuracy. It is of course possible to increase the order of accuracy of the solution algorithm; indeed the first-order method of solution is used mainly by way of introducing the subject in a classroom lecture; if a solution to the IVP (1.1) is desired for a practical purpose, the numerical analysis almost invariably makes use of a high-order method such as the Runge-Kutta.

As our purpose here is not to deliver a classroom lecture, but to develop a robust CFD code that will yield accurate solutions for practical problems, we ought to try to do better than to simply make use of Equations (3.78) to update our four fundamental variables. Indeed, as we shall show, Equations (3.78) are unsatisfactory not just from the point of view of accuracy, but also from the point of view of stability. Yes, despite the fact that we have chosen the time-step in accordance with the CFL criterion, and have put in artificial dissipation terms, we are still not quite out of the woods. The time-discretization scheme also has a role to play in the search for a stable algorithm.

The idea behind the Runge-Kutta method[61] is to compute the value of f(x,y) at several strategic points in the rectangle bounded by the points [x, x+h, y(x), y(x+h)] and combine them in such a way so as to increase the order of accuracy. Of course in order to 'establish' a point in this rectangle, it will be necessary to compute the unknown y(x+αh) (where α is a coefficient between 0 & 1) and this will have to be done from an equation similar to (1.3). In other words, a Runge-Kutta scheme involves several applications of formulae similar to (1.3) at each step - corresponding to the number of points in the rectangle; it has become customary to refer to each application as a 'stage' in the Runge-Kutta scheme. Each time a stage is added, the order of accuracy can be increased by 1 - *by appropriately choosing the points in the rectangle* - or equivalently - *by appropriately choosing the coefficients in the scheme.* Choosing the coefficients can be quite a cumbersome task; the interested reader could refer to the numerical analysis text book by Ralston and Rabinowitz (1978) for example. Let us just say here that for a particular number of stages, there are actually an infinite number of schemes having the maximum order of accuracy.

The best-known of the Runge-Kutta methods is the fourth-order scheme:

$$f_1 = f(x,y) \ / \ f_2 = f(x+\tfrac{1}{2}h,y+\tfrac{1}{2}hf_1) \ / \ f_3 = f(x+\tfrac{1}{2}h,y+\tfrac{1}{2}hf_2) \ / \ f_4 = f(x+h,y+hf_3)$$

$$y(x+h) \ = \ y(x) + h[\tfrac{1}{6}\{f_1+2f_2+2f_3+f_4\}]. \qquad \ldots (3.79)$$

---

[61] After the German mathematicians, Carl Runge (1856-1927) and Wilhelm Kutta (1867-1944).

Considering the accuracy of the method, the coefficients are astonishingly simple, and if a Runge-Kutta method is to be employed for the present problem (where 'x' corresponds to time and 'y' to our four fundamental variables), it would seem natural to use a formula analogous to (3.79). Strange as it may seem, our time-marching algorithm is - in one particular sense - actually simpler than the IVP (1.1)! This is because the quantities corresponding to f(x,y) are independent of the first argument, that is 'time'. If we omit the factor Δt in Equations (3.69) (which corresponds to the step h), then what remains is independent of 't' and the same is true of the artificial dissipation terms. Thus what we seek is a Runge-Kutta formula for the simpler IVP:

$$y' = f(y) \qquad \qquad \cdots \text{(3.80a)}$$
$$y(x_0) = y_0, \qquad \qquad \cdots \text{(3.80b)}$$

for which the Runge-Kutta formula (3.79) reduces to

$$f_1 = f(y) \ / \ f_2 = f(y+\tfrac{1}{2}hf_1) \ / \ f_3 = f(y+\tfrac{1}{2}hf_2) \ / \ f_4 = f(y+hf_3)$$
$$y(x+h) = y(x) + h[1/6\{f_1+2f_2+2f_3+f_4\}]. \qquad \cdots \text{(3.81)}$$

If we are to use an analogous method for our present problem, then 'y' will have to be replaced by the vector **U** [Equation (3.16), page 49], composed of the four fundamental variables, and f will take the form of Equations (3.69) (omitting the factor Δt) and the dissipation terms. Although this is certainly feasible, there is a drawback which must give us pause for thought. The quantities corresponding to f_1, f_2, f_3 and f_4 will naturally have to be stored at each point (or each cell) in the grid; thus we will require four additional arrays for each of our four fundamental variables, making 16 in all. This may be a worthwhile price to pay if the eventual solution is indeed fourth-order accurate, but this will not be the case because the space-discretization is not. Moreover accuracy in the time-discretization is in any case not of real significance because we are interested only in the final steady state.

Why then must we take the trouble of adopting Runge-Kutta methods if the accuracy of the results will in any case not correspond to the order of the method used? The answer is that the increase in the number of stages at each step has the effect of stabilizing the algorithm. The explicit central-differencing which was employed in the derivation of Equations (3.69) (page 72) has some inherently unstable features which can only be rectified by the use of a scheme with three or more stages. And if we are not particular about achieving third (or fourth) order accuracy, it will be possible to employ a three (or four) stage Runge Kutta scheme which does not require twelve (or sixteen) additional storage arrays.

Still in the context of the ODE (3.80), consider an m-stage Runge-Kutta scheme of the form:

$$y_1 = y(x) + \alpha_1 hf(y)$$
$$y_2 = y(x) + \alpha_2 hf(y_1)$$

$$\cdots$$
$$\cdots$$

$$y_{m-1} = y(x) + \alpha_{m-1} hf(y_{m-2})$$
$$y(x+h) = y_m = y(x) + \alpha_m hf(y_{m-1}). \qquad \cdots (3.82)$$

The coefficients $\alpha_1, \alpha_2 \ldots \alpha_m$ are not all arbitrary; in particular, α_m must be set equal to 1 for consistency, because the final formula simply represents a refinement of (1.3), where the function f has now been obtained to m^{th} order accuracy[62]. The big advantage in using a scheme such as (3.82) is that it is no longer necessary to store all the intermediate stages, as the eventual formula only involves the penultimate stage. Thus, for our purpose, it is certainly desirable to extend the scheme (3.82) rather than (3.81); we will still require one additional array for each of our four fundamental variables but that is all.

Thus, instead of (3.78), we will actually update the four fundamental variables using the algorithm:

$$U_1 = U(t) + \alpha_1(\Delta U + AU)$$
$$U_2 = U(t) + \alpha_2(\Delta U_1 + AU_1)$$

$$\cdots$$
$$\cdots$$

$$U_{m-1} = U(t) + \alpha_{m-1}(\Delta U_{m-2} + AU_{m-2})$$
$$U(t+\Delta t) = U_m = U(t) + \alpha_m(\Delta U_{m-1} + AU_{m-1}), \qquad \cdots (3.83)$$

where

$$U = \begin{bmatrix} \rho \\ \rho u \\ \rho v \\ \rho e \end{bmatrix} : \Delta U = \begin{bmatrix} \Delta \rho \\ \Delta \rho u \\ \Delta \rho v \\ \Delta \rho e \end{bmatrix} : AU = \begin{bmatrix} A\rho \\ A\rho u \\ A\rho v \\ A\rho e \end{bmatrix} \qquad \cdots (3.84)$$

The next decision to be taken is the number of stages to be used in the scheme and the choice of suitable coefficients. As regards the choice of coefficients, one natural way is to select them so as to increase the order of accuracy with each stage. Thus for a single-stage scheme, we must have $\alpha_1 = 1$; for a two-stage scheme, we must have $\alpha_1 = 1/2$, $\alpha_2 = 1$; for a three-stage scheme, $\alpha_1 = 1/3$, $\alpha_2 = 1/2$, $\alpha_3 = 1$; for a four-stage scheme, $\alpha_1 = 1/4$, $\alpha_2 = 1/3$, $\alpha_3 = 1/2$, $\alpha_4 = 1$; and so on. These values may be obtained by a straightforward Taylors series expansion; for example if we consider the two stage scheme - with reference to (3.81):

---

[62] This is equivalent to the condition in the general Runge–Kutta schemes that the eventual formula must make use of an increment that is a weighted average of the functions f_1, f_2, etc.

$$y(x+h) = y_2 = y(x) + \alpha_2 hf(y_1) = y(x) + \alpha_2 hy_1{}' \quad \text{[from (3.80a)]}$$
$$= y(x) + \alpha_2 hy{}'(x) + \alpha_2\alpha_1 h^2 y''(x). \qquad \ldots (3.85)$$

But by the second-order Taylor-series expansion:

$$y(x+h) = y(x) + hy{}'(x) + \tfrac{1}{2}hy''(x) \qquad \ldots (3.86)$$

so that we must have $\alpha_2 = 1$ and $\alpha_1 = \tfrac{1}{2}$. The coefficients of the higher-order schemes may be obtained in a similar manner.

However as explained earlier, accuracy in the time-discretization is not of primary importance and we could instead choose the coefficients so as to offer a wider scope for stability. This brings us once again to the subject of stability. In Section 5, we made use of the Von Neumann method, but here it is more convenient to employ the so-called 'descriptive method' [*ala* G.D.Smith (1965)] or 'matrix method' [*ala* C.Hirsch (1988), Chapter 10]. This method expresses the solution vector at time-level '$n+1$' in terms of the solution vector at time–level 'n', in the form

$$\mathbf{U}_{n+1} = [\mathbf{A}]\mathbf{U}_n + \mathbf{B} \qquad \ldots (3.87)$$

and seeks conditions on the matrix $[\mathbf{A}]$ that will ensure stability. Note that the vector $\mathbf{U}$ in (3.87) has no connection with the vector $\mathbf{U}$ in (3.84); its components are the solutions at all grid points at a particular time-level, and has precisely as many components as there are points on the grid. The main advantage of the descriptive over the Von Neumann approach is that it is more complete in its outlook; for example it takes into account boundary conditions (incorporated in the vector $\mathbf{B}$); the main disadvantage is that it is rarely an easy task to derive the matrix $[\mathbf{A}]$ and the vector $\mathbf{B}$ and carry out the stability analysis. However, without going into details, it is fairly obvious from (3.87) that if errors in the solution are not to grow, then the matrix $[\mathbf{A}]$ must be bounded and in fact must have a norm ≤ 1. This condition will be satisfied if all the eigenvalues of $[\mathbf{A}]$ have moduli ≤ 1.

For our present purpose, it is actually more convenient to express the scheme in the form:

$$[dU/dt]_{n+1} = [\mathbf{A_0}]\mathbf{U}_n + \mathbf{B_0}. \qquad \ldots (3.88)$$

Let us first suppose that the time-discretization is being carried out using a single-stage scheme. Then Equation (3.88) can be discretized as

$$\mathbf{U}_{n+1} = \mathbf{U}_n + \Delta t\{[\mathbf{A_0}]\mathbf{U}_n + \mathbf{B_0}\}. \qquad \ldots (3.89)$$

If we compare (3.87) and (3.89), we see that

$$[\mathbf{A}] = [\mathbf{I}] + \Delta t[\mathbf{A_0}] \qquad \ldots (3.90)$$

where [I] is the identity matrix. The advantage of (3.89) over (3.87) is that it is generally simpler to compute the eigenvalues of [A_0]. For example, in the case of a first-order scheme, it is fairly straightforward to show that forward differencing leads to an [A_0] with positive eigenvalues, backward differencing leads to an [A_0] with negative eigenvalues and central differencing leads to an [A_0] with eigenvalues that are purely imaginary [see Hirsch (1988), Chapter 10]. This last case will yield complex eigenvalues for [A] itself; moreover all these complex eigenvalues will have a real part equal to one and therefore a modulus greater than one. Thus central-differencing in a single-stage scheme will lead to instability.

Equation (3.90) may be alternatively expressed as a so-called 'amplification factor'

$$g(z) = 1 + z \qquad \qquad \ldots (3.91)$$

where z is a complex eigenvalue and g(z) - which is analogous to [A] in (3.86) - expresses the manner in which any errors in the solution may amplify from one time-step to the next. Clearly, for stability, we must have $|g(z)| \leq 1$, which in the case of Equation (3.91) yields a circle in the complex plane centred at (-1,0) with radius 1. Obviously eigenvalues that are purely imaginary will not lie within this circle and once again we see that one-stage central differencing leads to instability.

If we increase the number of stages, then the amplification factor g(z) will take the form of a polynomial in z whose order will be equal to the number of stages and whose coefficients will be governed by the coefficients α_1, α_2, etc. in the scheme. More explicitly, for an m-stage scheme:

$$g(z) = 1 + \alpha_m z + \alpha_m \alpha_{m-1} z^2 + \ldots + \alpha_m \alpha_{m-1} \ldots \alpha_1 z^m \qquad \ldots (3.92)$$

and, if the coefficients α_1, α_2, etc. are chosen so as to increase the order of accuracy at each stage (as explained on page 81) then

$$g(z) = 1 + z + (1/2!)z^2 + (1/3!)z^3 + (1/4!)z^4 + \ldots \qquad \ldots (3.93)$$

Let us try to work out the region in the complex plane where $|g(z)| \leq 1$, for different orders of accuracy. Actually the task is rather more difficult than may appear at first sight, because we must first separate g(z) into real and imaginary parts, and then square and add these parts, so that the order of the polynomial which must eventually be analysed will be twice that of g(z) itself. To start with, let us consider the relatively simple case of a two-stage scheme:

$$\begin{aligned} g(z) = g(x+iy) &= 1 + (x+iy) + \tfrac{1}{2}(x+iy)^2 \\ &= [1+x+\tfrac{1}{2}(x^2-y^2)] + i[y(1+x)] \end{aligned} \qquad \ldots (3.94)$$

and the condition $|g(z)| \leq 1$ takes the form:

$$[1 + x + \tfrac{1}{2}(x^2-y^2)]^2 + [y(1+x)]^2 \leq 1, \qquad \dots (3.95)$$

which may also be expressed as

$$(1+x)^2(1+y^2) + \tfrac{1}{4}[x^2-y^2][(x+2)^2-y^2] \leq 1. \qquad \dots (3.96)$$

The advantage of (3.96) over (3.95) is that it expresses the doubly symmetric nature of the function; it is obviously symmetric about the x-axis, but it is also symmetric about the vertical line $'x = -1'$, which may be seen by replacing x by its mirror–image about this line i.e $(-2-x)$. Furthermore, if $y = 0$, then (3.96) reduces to

$$(1+x)^2 + \tfrac{1}{4}x^2(x+2)^2 \leq 1, \qquad \dots (3.97)$$

which holds only if x lies in the interval $[-2,0]$. Again if $x = -1$:

$$\tfrac{1}{4}(1-y^2)^2 \leq 1, \qquad \dots (3.98)$$

which holds only if y lies in the interval $[-\sqrt{3},\sqrt{3}]$. Thus the set of points in the complex plane where (3.96) is satisfied forms an egg-shaped region[63] bounded by the four points $(-2,0)$, $(-1,-\sqrt{3})$, $(-1,\sqrt{3})$ and $(0,0)$. Just as in the case of the one-stage amplification factor (3.90), this stability region also does not contain any section of the imaginary axis: in fact this may be directly seen by setting $x = 0$ in (3.96); the left-hand side reduces to $1+\tfrac{1}{4}y^4$ which is always greater than 1 when $y \neq 0$. Thus a central-differencing scheme (for which the eigenvalues are purely imaginary) will once again lead to instability.

Let us now move on to the three-stage and four-stage schemes. In these cases the polynomials corresponding to (3.96) will be of sixth and eighth order in x and y - and the analysis becomes rather cumbersome to carry out. Instead we will develop a Fortran program to numerically generate the regions in the complex plane where the condition $|g(z)| \leq 1$ holds.

PROGRAM 3.5 (pages 85 to 87) is a Fortran routine for generating stability regions for Runge-Kutta schemes of different order. It divided into four parts. The main program, starts from the x-axis and march forward in small steps of Y (given by DY, which is fed in by the user) and computes the left and right boundaries of the stability region, by twice calling SUBROUTINE SOLVE with different initial conditions. As we approach the upper boundary of the stability region DY is reduced so as to be able to obtain this part of the curve accurately.

---

[63] Note that though the region is egg-shaped, it is *not* an ellipse; the boundary is a curve of the fourth order.

```
C... PROGRAM 3.5: GENERATION OF STABILITY REGIONS OF ORDER 1 TO 4
      COMMON M
      COMMON /LOG/ OVER
      COMMON /PL/ XL (4,-90:90),XR(4,-90:90),Y(4,-90:90),N(4)
      LOGICAL OVER
      READ (5,*) DYO,FR
      DO 40 M = 1,4
      DY = DY0
      OVER = .FALSE.
      XL0 = -2.0
      XR0 = 0.0
      I = 0
      Y(M,I) = 0.0
10    CALL SOLVE (XL0, 0.1,XL(M,I),Y(M,I))
      CALL SOLVE (XR0,-0.1,XR(M,I),Y(M,I))
      WRITE (6,50) I,Y(M,I),XL(M,I),XR(M,I)
      IF (ABS(XR(M,I)-XL(M,I)).LE.0.1) GO TO 30
      IF (OVER) GO TO 20
      I = I+1
      Y(M,I) = Y(M,I-1) + DY
      XL0 = XL(M,I-1)
      XR0 = XR(M,I-1)
      GO TO 10
20    DY = FR*DY
      Y(M,I) = Y(M,I-1) + DY
      OVER = .FALSE.
      GO TO 10
30    N(M) = I+1
      I = N(M)
      XL(M,I) = 0.5*(XL(M,I-1) + XR(M,I-1))
      XR(M,I) = XL(M,I)
      Y(M,I) = Y(M,I-1)
      DO 40 I = -1,-N(M),-1
      Y(M,I) = -Y(M,-I)
      XL(M,I) = XL(M,-I)
      XR(M,I) = XR(M,-I)
40    CONTINUE
      CALL PLOT
      STOP
50    FORMAT (I3,' : ',F6.4,' : ',F7.4,' : ',F7.4)
      END
C
      SUBROUTINE SOLVE (X0,STEP,X,Y)
      COMMON /LOG/ OVER
      LOGICAL OVER
      EPSLN = 0.00001
      X = X0
      CALL RESID (X,Y,RES)
      IF (ABS(RES).LE.EPSLN) RETURN
10    X1 = X
      RES1 = RES
```

```
        X = X + STEP
        CALL RESID (X,Y,RES)
        IF (ABS(RES).LE.EPSLN) RETURN
        IF (RES*RES1.LT.0) GO TO 30
        IF (ABS(RES).LT.ABS(RES1)) GO TO 10
        X = X1
        RES = RES1
20      X1 = X
        RES1 = RES
        X = X - STEP
        CALL RESID (X,Y,RES)
        IF (ABS(RES).LE.EPSLN) RETURN
        IF (RES*RES1.LT.0) GO TO 30
        IF (ABS(RES).LT.ABS(RES1)) GO TO 20
        GO TO 40
30      RATIO = RES / (RES-RES1)
        X2 = X1
        RES2 = RES1
        X1 = X
        RES1 = RES
        X = X1 - RATIO*(X1-X2)
        CALL RESID (X,Y,RES)
        IF (ABS(RES).LT.EPSLN) RETURN
        IF (RES*RES1.LT.0) GO TO 30
        X1 = X2
        RES1 = RES2
        GO TO 30
40      WRITE (6,50) Y
        OVER = .TRUE.
        RETURN
50      FORMAT (' NO SOLUTION EXISTS FOR Y = ',F6.4)
        END
C
        SUBROUTINE RESID (X,Y,RES)
        COMMON M
        X1 = X
        Y1 = Y
        XX = X1
        YY = Y1
        IF (M.EQ.1) GO TO 10
        X2 = X**2-Y**2
        Y2 = 2*X*Y
        XX = XX + X2/2
        YY = YY + Y2/2
        IF (M.EQ.2) GO TO 10
        X3 = X*(X**2-3*Y**2)
        Y3 = Y*(3*X**2-Y**2)
        XX = XX + X3/6
        YY = YY + Y3/6
        IF (M.EQ.3) GO TO 10
        X4 = X**4 - 6*X**2*Y**2 + Y**4
```

```
       Y4 = 4*X*Y*(X**2-Y**2)
       XX = XX + X4/24
       YY = YY + Y4/24
10     RES = (1.+XX)**2 + YY**2 - 1.0
       RETURN
       END
C
       SUBROUTINE PLOT
       COMMON /PL/ XL (4,-90:90),XR(4:-90:90),Y(4,-90:90),N(4)
       DIMENSION XX(0:180),YY(0:180)
       READ *,IO,B,XMIN,XMAX,YMIN,YMAX
       E = 1.-B
       CALL GROUTE (' ')
       CALL GOPEN
       CALL RORIEN(IO)
       CALL GRPSIZ(XS,YS)
       CALL GVPORT(XS*B,YS*B,XS*E,YS*E)
       CALL GWBOX(XS*E,XS*E,0.0)
       CALL GLIMIT(XMIN,XMAX,YMIN,YMAX,0.0,0.0)
       CALL GSCALE
       DO 30 M = 1,4
       N2 = 2*N(M)
       N21 = N2+1
       DO 10 I = 0,N2
       XX(I) = XL(M,I-N(M))
       YY(I) =  Y(M,I-N(M))
10     CONTINUE
       CALL BDIMX(1,XX,N21)
       CALL BLIVEC(YY,N21,0,0)
       DO 20 I = 0,N2
       XX(I) = XR(M,I-N(M))
       YY(I) = Y(M,I-N(M))
20     CONTINUE
       CALL BDIMX(1,XX,N21)
       CALL BLIVEC (YY,N21,0,0)
30     CONTINUE
       XX(0) = -3.0
       YY(0) = 0.0
       XX(1) = 1.0
       YY(1) = 0.0
       CALL BDIMX(1,XX,2)
       CALL BLIVEC(YY,2,0,0)
       XX(0) = 0.0
       YY(0) = -3.0
       XX(1) = 0.0
       YY(1) = 3.0
       CALL BDIMX (1,XX,2)
       CALL BLIVEC(YY,2,0,0)
       CALL GCLOSE
       RETURN
       END
```

In fact the program does not sense the upper boundary until it actually oversteps the mark, at which stage it will find no solution to the left and right boundaries. It then returns to the previous step and continues its way upwards with the reduced step-length, the reduction in step-size being repeated if need be. The procedure terminates when the left and right boundaries approach within a desired limit. At this stage, the lower half of the boundary is set by symmetry. After computing four such boundaries (for the schemes of first, second, third and fourth order, given by M = 1,2,3,4, respectively), the main program calls the plotting subroutine PLOT and then ceases execution.

SUBROUTINE SOLVE obtains the left and right boundaries of the stability region by the so-called method of false position; it begins by calling SUBROUTINE RESID (which provides the 'residue' of the equation $[g(z)]^2 - 1 = 0$; in other words, it simply computes the l.h.s) at a series of X values (for the given value of Y) until it obtains two neighbouring values of X where the residue changes sign; from here onwards, the value of X which yields a residue less than a given small number ε can be obtained by a continuous process of interpolation. If the subroutine is unable to find a solution - the condition for which is an increasing residue in both directions - this information will be conveyed to the main program by setting the logical variable OVER = .TRUE.

SUBROUTINE RESID is in a sense the key to the whole program as it does the actual computation of the amplification factor $g(z)$. It makes use of the formulae:

$$
\begin{array}{llll}
x_2 = x^2-y^2 & ; & y_2 = 2xy & \ldots \ (3.99a) \\
x_3 = x(x^2-3y^2) & ; & y_3 = y(3x^2-y^2) & \ldots \ (3.99b) \\
x_4 = x^4-6x^2y^2+y^4 & ; & y_4 = 4xy(x_2-y^2) & \ldots \ (3.99c)
\end{array}
$$

where (x_2,y_2), (x_3,y_3) and (x_4,y_4) denote the real and imaginary parts of z^2, z^3 and z^4 respectively. Depending upon the number of stages in the scheme, each of these quantities (divided by the appropriate coefficient) is successively added to the amplification factor; $[g(z)]^2-1$ is then computed and its value returned to the parent subroutine SOLVE.

Finally SUBROUTINE PLOT plots out the boundaries of the four stability regions corresponding to the four different orders of the scheme. Information is sent in the form of the arrays XL(M,I), XR(M,I) and Y(M,I), the former two providing the X-coordinates of the left and right boundaries of the four regions at the same Y-coordinate (the first argument M denotes the order of the scheme, the second argument I is a parameter defining a given Y-coordinate). Obviously the Fortran statements in this subroutine can be used only on a workstation with access to the UNIRAS graphics package; otherwise the subroutine will have to be modified.

The stability regions corresponding to the first, second, third and fourth-order schemes are depicted in Figure 3.10 (page 89).

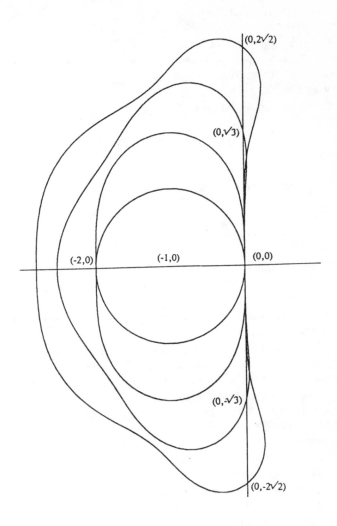

Figure 3.10   **Stability zones for Runge-Kutta schemes of different order**

It must be stressed that there is nothing new about Figure 3.10; it may be found, for example, in the eleventh chapter of Hirsch (1988) - and goes at least as far back as Lambert (1973)[64]. But then the purpose of this book is not so much to present new results but to provide the reader with a firm grounding on topics relevant to time-marching.

The most important point to be observed from Figure 3.10 is that the stabilty regions for the third-order and fourth-order schemes incorporate sections of the imaginary axis, so that central differencing becomes conditionally stable. In fact the boundaries of the third-order and fourth-order stability regions intercept the imaginary axis at the points $\pm\sqrt{3}$ and $\pm2\sqrt{2}$ respectively. This may be shown analytically by setting $z = iy$ in Equation (3.93) yielding

third-order: 　　　　　　$g(iy) = [1 - y^2/2] + i[y - y^3/6]$ 　　　... (3.100a)
fourth-order: 　　　　　　$g(iy) = [1 - y^2/2 + y^4/24] + i[y - y^3/6].$ 　　... (3.100b)

Squaring and adding the real and imaginary parts, we obtain:

third-order: 　　　　　　$|g(iy)|^2 = 1 - y^4/12 + y^6/36$ 　　　　... (3.101a)
fourth-order: 　　　　　　$|g(iy)|^2 = 1 - y^6/72 + y^8/576.$ 　　　... (3.101b)

The boundaries of the stability regions satisfy the equation $|g(iy)| = 1$, so it is clear from Equations (3.101) that the third-order and fourth-order boundaries intercept the imaginary axis at $\pm\sqrt{3}$ and $\pm2\sqrt{2}$ respectively.

What is the significance of these numbers? Well if we were dealing with a one-dimensional problem, then the condition for stability is that the Courant number - as defined on page 61 (first line) - must lie between these intercepts. At first sight, this appears to contradict the CFL condition, according to which the Courant number should not exceed 1. But this is not the case; the increase in the number of stages enlarges the domain of dependence of the discretized equation, so that it will now contain the domain of dependence of the differential equation for higher Courant numbers.

In the case of a two-dimensional problem (as in the present instance), we can now afford to multiply the right-hand-side of the inequality (3.56) (page 65) by a factor having a maximum value of $\sqrt{3}$ (for a third-order scheme) or $2\sqrt{2}$ (for a fourth-order scheme). Thus in practice both schemes are about equally good: the additional time required by the four-stage scheme is compensated for by the larger allowable time-step.

---

[64] Jameson and Baker (1983) depict the stability zones corresponding to various other schemes and also analyse the effect of dissipation.

Topic 9. Boundary conditions

Topics 5, 6, 7 and 8 provide all the material that is required to update the solution from one time-level to the next at all *interior* points (i.e cell-centres) of the flow domain. To summarise, suppose we are given the solution at a particular time-level (at both interior and boundary points). Then in order to obtain the solution at the next time-level, we must go through five basic steps[65]:

Step 1:
    Compute the local time-step. This is done by setting Δt in accordance with inequality (3.56) (page 65). Δt may then be multiplied by a factor of up to $\sqrt{3}$ if a three-stage scheme is being employed, or a factor of up to $2\sqrt{2}$ if a four-stage scheme is being employed.
Step 2:
    Compute the increments $\Delta \rho$, $\Delta \rho u$, $\Delta \rho v$, $\Delta \rho e$ from Equations (3.69) (page 72).
Step 3:
    Compute the artificial viscosities $A\rho$, $A\rho u$, $A\rho v$, $A\rho e$ using (3.72)...(3.77).
Step 4:
    Update the values of the four fundamental variables ρ, ρu, ρv, ρe at all interior points by adding the increments obtained in Step 2 and the A-V terms obtained in Step 3. Repeat the process for m stages in accordance with Equation (3.83) (page 81).
Step 5:
    Having computed the solution at the next time level at all interior points, apply the boundary conditions.

Steps 1 to 4 have been covered in Topics 5 to 8; it now remains to deal with Step 5. We can divide it into four 'sub-steps' according to the nature of the boundary under consideration: inlet, outlet, periodic, wall. Each set of boundary conditions is quite independent of the others and the boundaries may be operated upon in any order, though in the actual code, we start with the inlet, move on to the periodic and wall conditions and terminate with the outlet. In this discussion however, we will leave the wall boundary to the last, as it is the most difficult to deal with (in comparison to the other three).

Before describing the method of applying the boundary conditions at each of the four sub-steps, some general comments are in order. At all interior points, the emphasis was on the computation of the four fundamental variables ρ, ρu, ρv & ρe after which all other variables of interest could be obtained from purely algebraic formulae. But for the purpose of imposing the boundary conditions, it is more convenient to work with the straightforward ρ, u, v, e, p, t, h, q, c, etc.

---

[65] Obviously the sense in which we use the word 'step' should not be confused with 'time-step'. We could instead use the word 'stage' except that it could be similarly confused with 'Runge-Kutta stage'.

Our purpose is to derive formulae for specifying these variables at all boundaries. Some of these formulae may arise directly from the physical conditions of the problem, for example the prescribed total pressure and temperature at inlet and the tangency condition at the walls. In general, however, these physical conditions will not suffice to specify all the flow variables. They must be supplemented by so-called numerical boundary conditions which generally involve extrapolation from the interior of the domain. The totality of physical and numerical boundary conditions must suffice to make the problem 'well-posed' - that is - yield a unique solution.

The physical conditions relating to the problem will depend first-and-foremost on the nature of the freestream; thus if the freestream is supersonic, it would be quite meaningless to impose a condition at the outlet boundary as no information would propagate upstream. Remember that information is fed along characteristics and in the case of the Euler equations the characteristic directions are governed by the eigenvalues (3.24) (page 53) which are always positive - that is directed upstream - for supersonic flow. Thus in this instance, all conditions at outlet would have to be imposed numerically. In the present code, ρ, u, v and p are obtained by linear extrapolation from the interior of the domain; the temperature t may now be obtained from the equation of state (2.8) (page 26); the energy e and the enthalpy h may then be obtained from Equations (2.9 a,b).

However if the freestream is subsonic, then the outlet boundary also has a contribution to make; more specifically precisely one condition must be specified on this boundary because precisely one of the four eigenvalues (3.24) is negative. It is customary to specify the static pressure p. Then ρ, u and v must be computed by extrapolation and the other variables obtained as before from Equations (2.7) and (2.8).

The eigenvalues (3.24) also govern the specification of inlet boundary conditions; if the flow is supersonic, all four eigenvalues are positive and four physical boundary conditions must therefore be specified at inlet; if the flow is subsonic, this figure reduces to three. It is customary to specify total pressure, total temperature and inlet flow angle for both supersonic and subsonic flows - and to specify the inlet Mach number as well for supersonic flows. The present code has been written under the assumption that these are the physical boundary conditions provided, but it is not difficult to modify the code to deal with other given boundary conditions always providing the correct number of physical conditions are applied at the two boundaries depending upon the nature of the flow.

A special case arises if the flow at the inlet has both supersonic and subsonic features i.e is supersonic but with a subsonic axial component u. In this case, the eigenvalue $u-c$ is still negative and so the flow at inlet will still have some dependence on an outlet condition (static pressure) provided the characteristic whose direction is defined by this eigenvalue is not interrupted by the presence of a shock. If the user has reason to believe (eg from experimental evidence) that the flow passage contains a shock, then the inlet must be treated as essentially supersonic in nature. Moreover instead of specifying the inlet flow angle, it is customary to specify the inlet swirl velocity (i.e. the velocity perpendicular to the axial direction - the tangential velocity -which, for our two-dimensional problem is simply v). As in the subsonic case, the axial velocity (or alternatively the total velocity) must be obtained by extrapolation[66].

How are these conditions put into practice? In the code, the inlet boundary conditions are imposed in a subprogram entitled SUBROUTINE INLET. This subprogram begins by checking whether the freestream is subsonic or supersonic. If the former, the total velocity $q = \sqrt{(u^2+v^2)}$ is first obtained by extrapolation from the interior (*numerical b.c*). The temperature t is next obtained from the formula

$$t = T_0 - \tfrac{1}{2}q^2/c_p, \qquad \qquad \dots (3.102)$$

where T_0 is the total temperature (*physical b.c*) and c_p is the coefficient of pressure (set equal to 1005.0 at the start of the program). Next the speed of sound c is obtained from the formula

$$c = \sqrt{(\gamma R t)}, \qquad \qquad \dots (3.103)$$

where the ratio of specific heats γ and the gas constant R are set (equal to 1.4 and 287.0 respectively) at the start of the program. Next the Mach number M is obtained from the formula

$$M = q/c. \qquad \qquad \dots (3.104)$$

On the other hand, if the freestream is supersonic, the Mach number M is specified at inlet (*physical b.c*). The first step is to compute the temperature t from the formula (3.32b) (page 56). The speed of sound c may now be obtained as before from (3.103) but instead of solving for M in terms of q & c as in (3.104), the program now solves for q in terms of M and c:

$$q = Mc. \qquad \qquad \dots (3.105)$$

Thus regardless of whether the freestream is subsonic or supersonic, t, q and M have been obtained. The pressure p may now be obtained in terms of M and the total pressure P_0 (*physical b.c*) using formula (3.32a) (page 56). Density ρ follows from (3.33). Finally u & v are obtained from q and the given inlet flow angle or swirl velocity.

---

[66] The reason for specifying swirl velocity rather than inlet flow angle, for this particular case is not altogether clear to the author of the present work. From a numerical point of view, it ought not to make a great deal of difference, but the choice is probably based more on physical considerations.

Application of the periodic boundary conditions is quite straightforward (and is carried out in the subprogram entitled SUBROUTINE PERIOD). Two layers of cells are added on to the flow domain, one just below the lower periodic boundary and one just above the upper periodic boundary. The variables at the layer of cells below the lower periodic boundary are now set equal to their counterparts in the main flow domain below the upper periodic boundary. Similarly the variables at the layer of cells above the upper periodic boundary are set equal to their counterparts in the main flow domain above the lower periodic boundary.

We come now to the most difficult case of all: the wall boundary conditions. The basic difficulty arises from the fact that there is essentially just one physical b.c, governing the velocity components, the so-called tangency condition which states that the velocity vector at any point on the wall must be a tangent to its surface. Thus if f denotes the equation of the surface, the tangency condition states that

$$v = f'u. \qquad \ldots (3.106)$$

But this condition alone will not suffice to compute both u & v. Thus one of the two components will have to be obtained by a numerical b.c, that is extrapolation from the main flow domain. But which one?

At one time, when the author of this work was experimenting with different methods of imposing wall boundary conditions, he tried to avoid the use of numerical boundary conditions by supplementing (3.106) with one of the governing equations. Consider first the relatively simple case of incompressible flow. Then the continuity equation reduces to

$$u_x + v_y = 0. \qquad \ldots (3.107)$$

Can we use both (3.106) and (3.107) to solve for u and v and thereby avoid the need for a numerical b.c? Here is one approach which at first sight looks plausible. Assuming we are dealing with the lower boundary, $j = 0$, express $u_x(i,0)$ as a function of $u(i,0)$ and other values of u on the same *horizontal* line (in general this will involve interpolation between known values of u at interior grid points). Similarly express $v_y(i,0)$ in terms of $v(i,0)$ and other values of v on the same *vertical* line. Thus we have

$$u_x(i,0) = Au(i,0) + A' \qquad \ldots (3.108a)$$
$$v_y(i,0) = Bv(i,0) + B', \qquad \ldots (3.108b)$$

where the quantities A, A', B, B' are all known, for example if (3.108b) is obtained by first-order differencing, then $B = -1/\Delta y$ and $B' = v(i,1)/\Delta y$, where Δy is the distance between the points $(i,0)$ and $(i,1)$ assumed to be on the same vertical line. Now substituting (3.107) in (3.108), we may write

94

$$Au(i,0) + Bv(i,0) = C, \qquad \dots (3.109)$$

where $C = -(A'+B')$. Substituting the tangency boundary condition (3.106), this yields

$$u(i,0) = C / [A+Bf'] \qquad \dots (3.110a)$$
$$v(i,0) = f'C / [A+Bf']. \qquad \dots (3.110b)$$

In one sense, the use of formulae (3.110) is more satisfactory than the use of Equation (3.106) to obtain one of the velocity components and the other from a numerical extrapolatory formula, because no distinction is made between the method of computing u and v; moreover a higher order of accuracy can be obtained with a given number of interior points[67]. However there is a serious drawback to this approach which is not obvious at first. We will discuss this drawback shortly, but first let us see how the method may be extended to compressible flows.

Instead of complementing (3.106) with the continuity equation (3.107), we shall complement it with the steady form of the Euler equations:

$$\mathbf{F}_x + \mathbf{G}_y = 0 \qquad \dots (3.111)$$

where $\mathbf{F}$ and $\mathbf{G}$ are given by (3.16) (page 49). Then corresponding to (3.108), we have

$$\mathbf{F}_x(i,0) = A\mathbf{F}(i,0) + \mathbf{A}' \qquad \dots (3.112a)$$
$$\mathbf{G}_y(i,0) = B\mathbf{G}(i,0) + \mathbf{B}' \qquad \dots (3.112b)$$

and substitution of (3.111) yields

$$A\mathbf{F}(i,0) + B\mathbf{G}(i,0) = \mathbf{C} \qquad \dots (3.113)$$

corresponding to (3.109). $\mathbf{C}$ is a vector whose components consist of interior values of the components of $\mathbf{F}$ and $\mathbf{G}$. The vector equation (3.113) may be broken up into its component parts:

$$A(\rho u) \quad + B(\rho v) \quad = C_1 \qquad \dots (3.114a)$$
$$A(\rho u^2+p) + B(\rho uv) \quad = C_2 \qquad \dots (3.114b)$$
$$A(\rho uv) \quad + B(\rho v^2+p) = C_3 \qquad \dots (3.114c)$$
$$A(\rho uh) \quad + B(\rho vh) \quad = C_4. \qquad \dots (3.114d)$$

---

[67] Thus if u or v is obtained by numerical extrapolation, first-order accuracy would require two interior points, second-order accuracy three interior points, etc. On the other hand, each of formulae (3.108) requires one interior point for first-order accuracy, two interior points for second-order accuracy, etc.

From (3.114 a,b): $C_1u + Ap = C_2$ ... (3.115a)
From (3.114 a,c): $C_1v + Bp = C_3$. ... (3.115b)

The pressure p may be eliminated from Equations (3.115) to yield

$$C_1(Bu-Av) = BC_2-AC_3 \qquad \qquad ...(3.116)$$

and by substituting the tangency condition (3.106), we can solve for u and v. Now ρ can be got from (3.114a), p from (3.114 b or c) and h from (3.114d).

Although this method appears logical, it is seriously flawed. The 'bug' is the term $[A+Bf']$ which appears in the denominator in Equations (3.110)[68]. As a rule, if f' is positive, A & B are of opposite signs, whereas if f' is negative, A & B are of the same sign. This means that the term $[A+Bf']$ could well be zero at some points on the surface; indeed it actually *is* zero for several 'natural' grids. Consider, for example, a square grid imposed on domain bounded by a wall rising at an angle of 45° as in Figure 3.11a (page 97). Then $f' = 1$ and first-order differencing in x and y will yield

$$A = 1/\Delta x, \ A' = -u(i-1,0)/\Delta x \ ; \ B = -1/\Delta y, \ B = u(i,j+1)/\Delta y.$$

Since the grid is square, $\Delta x = \Delta y$, so that $A = -B$ and $[A+Bf'] = 0$. The situation is reversed in Figure 3.11b; here $f' = -1$, but $A = -1/\Delta x = B$ and once again we have $[A+Bf'] = 0$. Even when the wall forms the upper boundary as in Figure 3.11c and Figure 3.11d, the result is the same. In Figure 3.11c, $f' = -1$, $A = 1/\Delta x$, $B = 1/\Delta y$; in Figure 3.11d, $f' = 1$, $A = -1/\Delta x$, $B = 1/\Delta y$; in both cases $[A+Bf'] = 0$.

We may try and avoid this predicament by carefully setting up a grid such that the problem does not occur, but this is most unsatisfactory. Grid generation is difficult enough as an independent task - without the need to worry about extraneous factors. Moreover, for a configuration such as Figure 3.11 characterized by a 45° wall, it is most natural to use a square grid - and it is 'against the grain' to discard this grid simply because a particular formula is unworkable. We would be more inclined to discard the formula instead! The philosophy here is similar to that of testing a method on a basic problem; if it is unsatisfactory, the method is certainly not going to work on a more complex problem, though the complexities in the latter often have the effect of camouflaging the flaws in the method. And so, since the method of imposing the wall boundary condition described in the preceding pages failed to work on the basic grids depicted in Figures 3.11, it must reluctantly be discarded. Any attempt to complexify either the grid or the differencing formulae for u_x & v_y will only serve to confuse the issue but not solve the problem.

---

[68] Though not explicitly stated, this term also appears in the denominator of formulae in the compressible case.

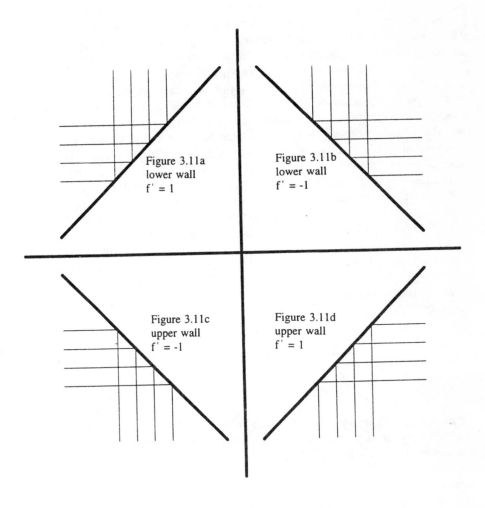

Figure 3.11 a,b,c & d   **Model square-grids on 45° walls**
(to test the practicability of a method of imposing
wall boundary conditions)

At this stage the reader may well ask whether any purpose was served in this detailed description of a plausible method of imposing the wall boundary conditions, when the method is not in fact going to be used in the final program. Before answering this question, the author must emphasize that this manual is directed not merely at persons who wish to use the program mechanically, but also to (in fact chiefly to) serious students of research for whom any relevant subject could be of importance in their future efforts to improve upon an existing piece of work. Thus the method of imposing wall boundary conditions described on pages 94 to 96, though inoperable, has an important lesson to teach. Why does the method break down? Simply because we are, in a sense, attempting to determine quantities that are indeterminate. Yes, surprising though it may seem, the velocity components u and v are indeterminate on a wall boundary[69] in the sense that they may assume any values satisfying the tangency condition (3.106) and will have no effect on the solution in the main flowfield. This is because the Euler equations in conjunction with the tangency boundary condition - which may be restated as the condition that there is no component of velocity perpendicular to the wall - have the effect of depriving the main flowfield of any information from the wall boundary with the exception of pressure values.

Let us see just why this is so. For convenience (but without loss of generality), let us assume that the wall is horizontal, so that $v = 0$. We shall now show that the boundary value of u has no effect on the value of u at any point in the interior of the flow domain however close it may be to the boundary.

Consider the steady form of the Euler equations:

$$
\begin{aligned}
(\rho u)_x &\quad + (\rho v)_y &&= 0 & \quad \ldots \text{(3.117a)} \\
(\rho u^2 + p)_x &\quad + (\rho u v)_y &&= 0 & \quad \ldots \text{(3.117b)} \\
(\rho u v)_x &\quad + (\rho v^2 + p)_y &&= 0 & \quad \ldots \text{(3.117c)} \\
(\rho u h)_x &\quad + (\rho v h)_y &&= 0. & \quad \ldots \text{(3.117d)}
\end{aligned}
$$

Obviously if a boundary value of u is to have any influence on an interior point, this can only arise through a y-derivative term involving u. Thus only the momentum equation in the x-direction need be taken into consideration. Its y-derivative term is $\rho u v$. But because of the presence of v, this term is always zero on the boundary regardless of the value of u. In other words we can assign any boundary value we wish to u without affecting the mainstream solution.

In the preceding argument, we assumed the wall to be horizontal, but this is of no significance to the proof. In fact if the wall is not horizontal, the identical argument applies except that instead of u & v, we have the tangential and normal components of velocity at the wall. Then the normal component is zero and the tangential component has no effect on the solution on the interior. In this instance, the tangential component is simply the total velocity q, and since q is indeterminate, so are its horizontal and vertical components u and v.

---

[69] Unless the wall is horizontal, in which case $v = 0$, or vertical, in which case $u = 0$.

Does this mean that the Euler equations cannot be solved on a wall surface? Well not exactly. After all, the equations can be solved at any interior point arbitrarily close to the wall surface. By continuity, the wall values of the velocity components must be equal to their limiting values as one approaches the wall from the interior. Therefore, for practical purposes, the wall velocity components may be found by extrapolating from the interior; and to ensure satisfaction of the tangency boundary condition (3.106), it is preferable to extrapolate the total velocity q and then compute u and v by separating q into its components.

Let us now summarise. At any stage of the solution process, the wall values of u and v may be obtained by extrapolating the total velocity q from the interior and applying the tangency boundary condition. However as they have no effect on the course of the solution procedure, there is no particular need to compute them at every time step and a little CPU time may be saved (but only very little!) by computing these wall values only after convergence has been attained. In the actual program, the wall values of u and v are computed at each time step (and are used in the evaluation of artificial viscosity terms) but the reader may dispense with this if he wishes.

This leads us to an important philosophical question? If the wall values of u and v have no influence on the interior solution, then in what way does the wall influence the flow? Obviously it must! The answer is that the influence automatically arises from the tangency condition, in other words the condition that the component of flow normal to the wall is zero. Consider Equations (3.69). The four fundamental variables ρ, ρu, ρv and ρe are being updated at each grid cell. The right-hand-sides of Equations (3.69) involve the mass flows entering and leaving the four faces of each cell. If we are dealing with a cell adjacent to a wall boundary, then no mass flow can enter or leave that wall. By applying this simple condition on Equations (3.69) at all boundary cells, we can correctly impose the influence of the wall boundaries.

Equations (3.69) provide additional confirmation that the wall values of u and v have no influence on the interior solution. In fact u & v (and h) only appear as coefficients of Q_1 and Q_2, the mass flows in the x and y directions. At a wall Q_2 will be zero and the associated value of u, v or h is therefore immaterial.

One other variable which appears in Equations (3.69) is the pressure p - and, significantly, it does not appear as a coefficient of Q_1 or Q_2. This means that the wall value of p does have some influence on the interior solution, and consequently the computation of the wall values of p is of considerably more importance than the computation of the other flow variables at the wall. It is of course possible to obtain the wall values of p by extrapolation from the interior, but let us try and do better.

It so happens that for inviscid flow, the pressure gradient normal to a wall surface can be expressed in terms of the tangential velocity and the radius of curvature at the point under consideration. We have already had an occasion to refer to this relationship - Equation (3.10) (page 44); it is now time to prove that it holds.

99

Consider the momentum equation in the y-direction: Equation (3.117c). It can be expanded in the form:

$$(\rho u)_x v + (\rho u)v_x + (\rho v)_y v + (\rho v)v_y + p_y = 0. \qquad \ldots (3.118)$$

Without loss of generality, we may rotate the coordinate axes in such a way that the normal to the wall surface at the point under consideration is directed vertically upwards. The situation is depicted in Figure 3.12 (page 101) [70]. At the point (x_c, y_c), the normal velocity v is zero (tangency boundary condition) and (3.118) reduces to

$$p_y = -(\rho u)v_x. \qquad \ldots (3.119)$$

What do we know about v_x? To start with, let us break it up into the components

$$v_x = v_\xi \xi_x + v_\eta \eta_x \qquad \ldots (3.120)$$

where ξ and η are orthogonal curvilinear coordinates, the former defining arc-length along the streamlines. Of course we do not know the streamlines in the main flowfield, but one of them is the wall surface itself and that is the one that matters.

Now the second of the two terms on the right hand side of (3.120) is zero. This can be intuitively felt from Figure 3.12, because the values of η along the straight line '$y=y_c$' are the distances from the wall surface along the transverse grid lines, so that the graph of η with respect to x will assume the form of a curve that is approximately an inverted image of the wall surface. Its derivative will be negative to the left of (x_c, y_c), positive to the right of (x_c, y_c) and zero at (x_c, y_c) itself. We can also prove this by making use of the relation '$\eta_x = -y_\xi/J$' [see Equation (3.7c) (page 43)]. The streamline passing through the point (x_c, y_c) is the wall surface itself, and if its equation is given by f(x), then

$$y_\xi = f'(x) \, x_\xi \qquad \ldots (3.121)$$

and $f'(x)$ is zero at the point (x_c, y_c). Thus (3.120) reduces to

$$v_x = v_\xi \xi_x \qquad \ldots (3.122)$$

Because ξ defines the arc length along the wall surface, we have

$$\xi = \int [1+f'(x)]dx \qquad \ldots (3.123)$$

---

[70] This figure is similar to Figure 3.6b (page 47). Our purpose there was to show how the radius of curvature could be computed; our purpose here is to show how it relates to the normal pressure gradient.

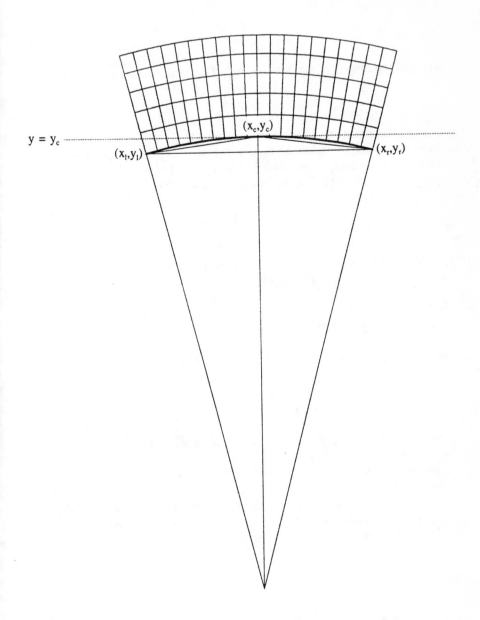

Figure 3.12   Derivation of a wall boundary condition for the pressure-gradient in terms of radius of curvature

where the limits of integration range from the the x-coordinate of the point where ξ is set at zero (say the leading edge) to the x-coordinate of the point under consideration. From (3.123):

$$\xi_x = 1+f'(x) \qquad \ldots (3.124)$$

and is simply equal to 1 at the point (x_c,y_c). Therefore (3.122) reduces to

$$v_x = v_\xi \qquad \ldots (3.125)$$

It remains to determine v_ξ. For this we will need to know v at points to the left or right of (x_c,y_c) and consider the limit as the point (x_c,y_c) is approached. For completeness, let us evaluate v at two points (x_l,y_l) and (x_r,y_r) one to the left and the other to the right of (x_c,y_c). Referring to Figure (3.12), we have

$$u(x_l,y_l) = q(x_l,y_l) \cos \theta \qquad \ldots (3.126a)$$
$$v(x_l,y_l) = q(x_l,y_l) \sin \theta \qquad \ldots (3.126b)$$

where q is the total velocity directed along the tangent to the wall surface, and θ is the angle made by the tangent with the x-axis. Differentiating with respect to x:

$$v_\xi(x_l,y_l) = q\cos\theta[d\theta/d\xi] + q_\xi\sin\theta = u[d\theta/d\xi] + q_\xi\sin\theta. \qquad \ldots (3.127)$$

On the other hand, at (x_r,y_r), the total velocity is directed downwards so that

$$u(x_r,y_r) = q(x_r,y_r) \cos \theta \qquad \ldots (3.128a)$$
$$v(x_r,y_r) = -q(x_r,y_r) \cos \theta \qquad \ldots (3.128b)$$

and

$$v_\xi(x_r,y_r) = q\cos\theta[-d\theta/d\xi] - q_\xi\sin\theta = u[-d\theta/d\xi] - q_\xi\sin\theta. \qquad \ldots (3.129)$$

It is now time to introduce into the analysis the 'radius of curvature' R_c[71]. The 'curvature' of a geometric curve is defined as its rate of change of direction with respect to arc length. Thus a circle with radius equal to 1 has a curvature also equal to 1 because the length of any arc is precisely equal to the angle subtended at the centre and hence to the difference in tangential direction at its

---

[71] The first published work on the 'radius of curvature' was by the Swiss mathematician Jakob Bernoulli (1654-1705), who in an article in the *Acta Eruditorium* in 1694 provided formulae for expressing it in Cartesian coordinates. However the subject was also known to Newton; he discussed it in a Latin work entitled *Methods of Fluxion* in 1671, but it remained unpublished until 1736.

Incidentally Bernoulli is also the mathematician credited with introducing the word 'integral' (*Acta Eruditorium*, 1690). These and other details may be found in the book: *History of Mathematics* by Florian Cajori (1919).

extremes. It can be verified that the radius of any circle is the reciprocal of the absolute value of its curvature and it is natural to extend this definition to any curve.

In Figure 3.12, θ defines the tangential direction of the wall surface and ξ is a measure of the arc length. Thus $d\theta/d\xi$ denotes curvature and $|d\theta/d\xi|$ the inverse radius of curvature. To the left of the point (x_c,y_c), θ decreases to zero; thus $d\theta/d\xi$ is negative and equal to $-1/R_c$. To the right of the point (x_c,y_c), θ increases from zero; thus $d\theta/d\xi$ is positive and equal to $1/R_c$. Substitution in (3.127) and (3.129) yields

$$v_\xi(x_l,y_l) = -u(x_l,y_l)/R_c + q_\xi\sin\theta \qquad \dots \text{(3.130a)}$$
$$v_\xi(x_r,y_r) = -u(x_r,y_r)/R_c - q_\xi\sin\theta. \qquad \dots \text{(3.130b)}$$

Now let θ tend to zero. The second term in both equations vanish leaving us with

$$v_\xi(x_c,y_c) = -u(x_c,y_c)/R_c \qquad \dots \text{(3.131)}$$

Substitution of (3.131) and (3.125) into (3.119) yields the desired relationship for the pressure gradient:

$$p_y = \rho u^2/R_c. \qquad \dots \text{(3.132)}$$

Equation (3.132) has been derived under the assumption that the wall surface was horizontal at the point under consideration so that the normal was directed vertically upwards. More generally, if 'n' denotes the outward normal at any point on the wall surface, then in Equation (3.132) we must substitute 'n' for 'y' and 'q' for 'u', that is

$$p_n = \rho q^2/R_c \qquad \dots \text{(3.133)}$$

which is precisely the same as Equation (3.10) (page 44) because on a wall surface, the total velocity and the tangential velocity are the same.

The reader may recall that the radii of curvature at all grid points on the wall surface are computed at the start of the program and stored for future use (see Topic 2). The total velocity q may be obtained by extrapolation from the interior values in the flowfield[72]. Thus p_n may be obtained at each cell face on the wall

[72] On reading this statement, the reader may quite naturally raise the following objection: The main objective in deriving Equation (3.133) was to provide a formula which would enable the wall pressure to be determined without having to resort to the inaccuracies of extrapolation. Would not this objective be defeated if we now extrapolate for q? The answer is that any inaccuracy in this extrapolation for q would not be serious, because the term is being multiplied by the inverse radius of curvature, a small quantity. Indeed, several programs make use of the approximate boundary condition $p_n = 0$ (which will hold precisely for a flat surface) and obtain reasonably satisfactory results. Obviously it is better to set p_n equal to an approximate value

surface, and making use of the known value of p at the centre of the cell (and perhaps a neighbouring cell) we can obtain the value of p at the wall.

This essentially completes our discussion of the different methods of applying boundary conditions. In the actual program, it takes up five subroutines: INLET, OUTLET, PERIOD, FLATWALL and CURWALL. The first three are self-explanatory. SUBROUTINE FLATWALL applies the wall boundary conditions on the assumption that the surface is flat, in which case Equation (3.133) reduces to $p_n = 0$. The user is expected to specify this in the input; the program will then skip the subroutine for computing radii of curvature. SUBROUTINE CURWALL more generally applies the ideas discussed above.

Topic 10. General programming details

We have now completed our analysis of all the basic elements which go to make up the Euler solver. The final section in this chapter explains how all these elements are synthesized into the Fortran program that is listed in the next chapter.

In developing a Fortran program, the first step must be to set up the arrays - or COMMON BLOCKS. Let us review the quantities which require storage. First and foremost we must store the grid. Thus far we have been referring to the grid coordinates as X-Y (X horizontal, Y vertical), but in the actual program the coordinates employed are X and Z with X vertical and Z horizontal. The advantage of this formulation is that when the solver is extended to three dimensions, we may use a Y-axis which is parametrized *upwards* (as is natural) whereas if we employ the traditional X-horizontal Y-vertical 2-D coordinate system, then the extension to three dimensions would necessitate a Z-axis parametrized *downwards* (in keeping with the right-hand corkscrew convention of 3-D coordinate geometry). For this reason, it has now become fairly standard practice for 3-D CFD solvers to use a coordinate system in which Z is directed from west to east, X is directed from south to north, and Y is directed from bottom to top. In effect this makes Z the most important coordinate as the flow is conventionally modelled on a freestream that moves from west to east. The various X-Y planes form 'slabs' perpendicular to the main flow direction[73].

When writing his code, Cheng decided to adhere to the practice of the Z-oriented coordinate system for both his 2-D & 3-D solvers. This practice is also followed in the present work. Thus the coordinates of our system are X & Z.

---

of the right hand side of Equation (3.133) than to set it to zero.

[73] The use of the word 'slab' to describe a surface perpendicular to the main flow direction appears to have originated with the Imperial College team that developed the commercial CFD program PHOENICS. In this program - which is based on the pressure-correction technique of Patankar and Spalding (1972) - the solution at each iterative step is carried out slab by slab from leading to trailing edge.

These coordinates are specified at grid-points (rather than cell centres), the indices ranging from I = 0 at the lower boundary to I = M at the upper boundary and from K = 0 at the inlet to K = N at the outlet: see Figure 3.13a (page 107). These coordinates are stored in our first common block:

COMMON /BLK1/ X(0:M,0:N), Z(0:M,0:N).

Thus the total number of points in the grid is (M+1) x (N+1). Note however that the total number of cells in the grid depicted in Figure 3.13a is simply M x N. The cell centres on the lower boundary have the index number I = 1, those on the upper boundary have the index number I = M, those at inlet have the index number K = 1 and those at outlet have the index number K = N.

In our second common block, we store the four projective-lengths which were discussed in Topic 2. The nomenclature that Cheng followed here (and which is adhered to in the present work) is based on a forty-year-old paper by Wu (1952), in its time a landmark in the numerical computation of 3-D turbomachinery flows. Wu divided the flow domain into three sets of mutually orthogonal 'stream surfaces' S1, S2 and S3, and solved the governing equations as a 2-D problem on each set of surfaces one at a time. With the advent of Euler and Navier-Stokes flow solvers, his method is now outdated, but much of his terminology is still in use. In Cheng's 3-D solver, the S1 surfaces stretch from wall to wall (i.e pressure to suction surface), the S2 surfaces from hub to shroud and the S3 surfaces from inlet to outlet (these are approximately the X, Y and Z directions in the Cartesian system). In the 2-D solver, we will ignore the changes from hub to shroud and thus only the S1 and S3 surfaces (now reduced to curves) need be taken into consideration. The S1 curves are quasi-streamlines marked by a constant value of the index I; the S3 curves are quasi-equipotential lines marked by a constant value of the index K. Thus the south and north faces of any particular cell in our grid are S1 curves, the west and east faces are S3 curves. We shall denote by S1X the projective length of an S1 face along the plane 'X = constant'; similar definitions apply to S1Z, S3X and S3Z. Note that the arrays S1X and S1Z, being associated with the south-north faces of each cell, will utilize an I-index ranging from 0 to M, but a K-index ranging from 1 to N. On the other hand, the arrays S3X and S3Z will use an I-index from 1 to M but a K-index from 0 to N. Thus we may write our second common block as:

COMMON /BLK2/ S1X(0:M,N), S1Z(0:M,N), S3X(M,0:N), S3Z(M,0:N).

In our third common block, we store just one array - the cell areas (which will be computed in accordance with Equation 3.5b (page 40). In keeping with current CFD jargon, we shall actually refer to them as 'cell volumes', despite the 2-D setting (see footnote 54, page 68). This array will require precisely as many components as the grid cells depicted in Figure 3.13a, thus its I-index will simply range from 1 to M and its K-index will simply range from 1 to N:

COMMON /BLK3/ VOL(M,N).

The next four common blocks will be reserved for our four fundamental variables. Because the flow variables will be stored at cell centres, it may seem that each such array will require M x N elements, but in fact the number will be greater, because an additional layer of cells will be fixed alongside each boundary to facilitate the application of boundary conditions. In fact the main flow variables will stored at cell-centres on the grid depicted in Figure 3.13b, where the cells on the lower boundary have the index number I = 0, those on the upper boundary have the index number M1 (= M+1), those at inlet have the index number K = 0 and those at outlet have the index number N1 (= N+1).

The first of our four fundamental variables is the density ρ which we shall designate as RHO. In addition we will require an array for temporary storage of the density-values as they change during the various stages of the Runge-Kutta scheme. This will be designated as RHOT. Then there is the increment $\Delta\rho$ which we shall denote as DRHO. Putting these together, our fourth common block will appear as:

COMMON /BLK4/ RHO(0:M1,0:N1), RHOT(0:M1,0:N1), DHRO(M,N).

Note that DRHO only requires storage on the main grid, not on the boundaries.

In a similar way we can write the common blocks for our other three fundamental variables. However because we have changed the nomenclature of our coordinate system to X-Z, we must correspondingly change the nomenclature of our velocity components. We could use U & W, but in fact Cheng designated the components as VX & VZ, so we shall adhere to this. The energy will simply be denoted by E, so our four fundamental variables can be designated by the array names RHO, RVX, RVZ and RE, and our next three common blocks take the form:

COMMON /BLK5/ RVX(0:M1,0:N1), RVXT(0:M1,0:N1), DRVX(M,N).
COMMON /BLK6/ RVZ(0:M1,0:N1), RVZT(0:M1,0:N1), DRVZ(M,N).
COMMON /BLK7/ RE(0:M1,0:N1), RET(0:M1,0:N1), DRE(M,N).

In our eighth common block, we will store the pressure P and the quantity RH which is the product of the density and enthalpy. The two variables share a common feature; P is required in the updating procedure of the momentum equations; RH is required in the updating procedure of the energy equation. The two velocity components VX and VZ will be stored in Block-9, the thermodynamic variables (energy E, enthalpy H, and temperature T) in Block-10. All these quantities are associated with cell-centres rather than cell-faces, and must be computed on the main flow domain as well as on the boundaries; thus their array indices will range from 0 to M1 and 0 to N1. So we have:

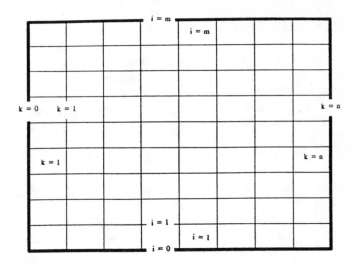

Figure 3.13a   The *interior* grid
(where the main flow equations will be solved)

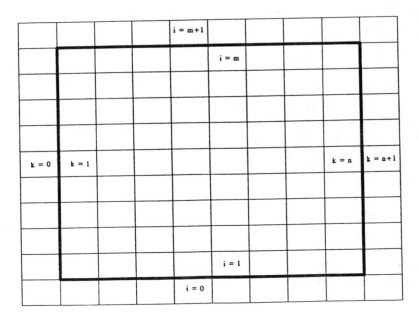

Figure 3.13b   The *entire* grid
(the flow in the cells outside the interior grid will be
updated by boundary conditions)

COMMON /BLK8/ P(0:M1,0:N1), RH(0:M1,0:N1)
COMMON /BLK9/ VX(0:M1,0:N1), VZ(0:M1,0:N1)
COMMON /BLK10/ E(0:M1,0:N1), H(0:M1,0:N1), T(0:M1,0:N1).

In our eleventh block, we store the velocity fluxes Q1 and Q3 [74]. Note that these quantities are associated with cell-faces rather than cell-centres, Q1 being associated with the south-north faces and Q3 with the east-west faces. Thus the array indices of Q1 will be the same as that of S1X & S1Z (Block 2) and the array indices of Q3 will be the same as that of S3X & S3Z, giving:

COMMON /BLK11/ Q1(0:M,N), Q3(M,0:N).

In our twelfth block, we store the speed of sound C and the local time step DT. The two quantities are intimately connected, because the speed of sound determines the size of the maximum eigenvalues, which in term determines the local time step in accordance with the CFL condition: see Equations (3.60) and (3.61). Both C & DT need be computed only on the main grid, indices 1-M, 1-N:

COMMON /BLK12/ C(M,N), DT(M,N).

All arrays relating to artificial viscosity will be stored in two special blocks designated /ART1/ and /ART2/. The former stores the coefficients AX1, AX3, AZ1 and AZ3 (see Equation 3.75, page 78) and the latter the actual artificial viscosity contributions to the four fundamental variables designated here as AVRHO, AVRVX, AVRVZ and AVRE (computed in accordance with Equation 3.72). The coefficients AX1 and AX3 are associated with the south-north faces of each cell (0 to M, 1 to N), the coefficients AZ1 and AZ3 with the east-west faces of each cell (1 to M, 0 to N), and the artificial viscosity contributions AVRHO, AVRVX, AVRVZ and AVRE with the cell centres in the main grid (1-M, 1-N). Therefore we may write:

COMMON /ART1/ AX1(0:M,N), AX3(0:M,N), AZ1(M,0:N), AZ3(M,0:N)
COMMON /ART2/ AVRHO(M,N),AVRVX(M,N),AVRVZ(M,N),AVRE(M,N).

This completes the basic storage requirements. The program contains a few other 'common blocks' which store some relevant constants and some information required in the application of the boundary conditions (for example the radii of curvature at different grid points at the two wall surfaces). The contents of these common blocks will be defined in the comment statements in the Fortran listing in the next chapter.

---

[74] Q1 & Q3 are slightly different to Q_1 & Q_3 defined in Topic 6. Here they are velocity (or volumetric) fluxes; there they were mass fluxes. Cheng made use of velocity fluxes in his program, and the present author found it convenient to keep his usage intact.

Let us now move on to the program proper. It begins by opening relevant input and output files: eight in all. File No 1 is the grid file, No 2 is the operating-conditions file, No 3 is the initial-guess file. All these three are input files; the user is expected to generate his own grid and initial guess, though some guidelines have been provided in Topics 1 and 4 of this chapter. The next five files are output. File No 4 ('areas.dat') prints out cell areas and projective lengths, No 7 ('history.dat') prints out iterative history at a selected monitor point, No 8 ('result.dat') prints out the main solution, No 9 ('graphin.dat') prints out selected values of the Mach number for purpose of input to a graphics program, and No 10 ('continue.dat') prints out the most recent iterative step onto a file which may subsequently be read if a continuation run is desired. Note that File Nos 5 & 6 have been skipped as these are generally reserved for the reader and printer.

Having opened the relevant files, the program begins to read the input data starting with the grid from File No 1. This file also provides four key numbers, the indices IM & KM (denoting the grid size in the two directions[75]), and KLE & KTE (the indices of the points at leading and trailing edge). LEVEL is an index specifying whether the wall is flat or curved; if the former, the program can save time by skipping the computations of the inverse radius of curvature and making use of the wall boundary condition $\partial p/\partial n = 0$.

Now we come to the operating-conditions input file. ISUPER and ISUBAX are indices specifying whether the flow is wholly supersonic, or supersonic with a subsonic axial component; if both are set to zero, the program will assume wholly subsonic flow. Next the program reads the boundary conditions PT (total pressure at inlet), TT (total temperature at inlet), etc.: for further details, see the comment statements in the program. TIMFAC, the parameter for varying the time-step must be set $\leq\sqrt{3}$ for a three-stage scheme and $\leq 2\sqrt{2}$ for a four-stage scheme in accordance with the analysis of Topic 8. The artificial viscosity coefficients A1 and A3 are generally set to 0.25 and 0.008, but the user can play around with these figures to try to get optimum results. MSTAGE, the number of stages of the Runge-Kutta scheme is generally set to either 3 or 4; a figure less than 3 would lead to instability as explained in Topic 8. Guidelines for the choice of coefficients CF(I) were discussed in Topic 8, but here again the user can experiment. Then follow the printing and convergence criteria, which should be self-explanatory.

The last input data item in the 'operating-conditions' file is the index INIT. If it is set equal to zero, the program will read the initial values of RHO, VX, VZ and P from File No 3 (the 'initial-guess file'). If it is set equal to 1, the reading is done from File No 10 (the 'continuation file'), which of course assumes that a first run has already been performed and the present run is a continuation.

---

[75] IM and KM must of course be less than or equal to M and N - which appear in the common blocks, and which must be parametrized at the start of the program. Alternatively M and N must be set at values $\geq$ IM and KM respectively.

Note that in both cases, the variables are read only in the main grid (1 to IM, 1 to KM) and not on the boundaries.

SUBROUTINE CHECK (which is called at this point) merely prints out the input data on the screen so that the user may be satisfied it has been specified correctly. After defining the fundamental constants c_p, c_v, γ and R, the program now calls SUBROUTINE GEOM which computes the quantities S1X, S1Z, S3X and S3Z in accordance with the formulae derived in Topic 2. If the wall boundary is curved, SUBROUTINE RADIUS is also called to calculate the inverse radii of curvature.

Next the thermodynamic variables (temperature T, enthalpy H and energy E) are initialized using Equations (2.8) & (2.9) (page 26). Note that this is done on the main grid *before* the boundary conditions are applied. However the 'running variables' RVX, RVZ, RH and RE are initialized on the complete grid (interior and boundary) *after* the boundary conditions are applied.

The main marching process commences by updating the time-step indicator ICOUNT. The fundamental variables RHO, RVX, RVZ and RE are now placed in temporary storage RHOT, RVXT, RVZT and RET (the latter will change at each stage of the Runge-Kutta scheme). After computing the fluxes Q1 and Q2, the speed of sound C $(=\sqrt{\gamma RT})$ and the local time step DT[76], the main program calls the subroutines ARTCF and ARTVIS which together compute the artificial viscosity contributions in accordance with the analysis of Topic 6. Note that these contributions are not updated at each stage of the Runge-Kutta scheme as this would be unnecessarily time-consuming. Next the main program calls the updating subroutines DELRHO, DELRVX, DELRVZ and DELRE which compute the increments DRHO, DRVX, DRVZ and DRE in accordance with Equations (3.69) (page 72). After each such cycle, the temporary values of the fundamental variables RHOT, RVXT, RVZT and RET are updated in SUBROUTINE CORRECTION in accordance with Equation (3.83) (page 81); moreover it will also be necessary to update VX, VZ, E, T, P, H and RH as all these variables will have an effect on the computations at the later stages of the Runge-Kutta scheme. Note that though VX and VZ are not directly involved in the sunbroutines DELRHO, DELRVX, DELRVZ and DELRE, they are used in the computation of the fluxes Q1 and Q2 which *are* involved in the above subroutines. SUBROUTINE FLUX is therefore called at each stage of the Runge-Kutta scheme.

The six subroutines DELRHO, DELRVX, DELRVZ, DELRE, CORRECTION and FLUX are called precisely as many times as the number of stages in the Runge-Kutta scheme. At the end of the final stage, the boundary conditions are applied and the four fundamental variables RHO, RVX, RVZ & RE are updated

---

[76] This must be done in accordance with inequality (3.61). However the term $u\Delta y$ and $v\Delta x$ in that equation must be replaced by the appropriate velocity fluxes entering the cell-faces. This involves the projective-lengths S1X, S1Z, S3X and S3Z. The reader should have no difficulty following the Fortran steps on page 117.

by setting them to their temporary analogues RHOT, RVXT, RVZT & RET. Note however that RHO must be updated in the main (i.e interior) grid *before* the application of the boundary conditions. The other fundamental variables are updated on the complete grid *after* the application of the boundary conditions.

The time-step cycle is now repeated until convergence to a desired accuracy is obtained. Convergence is tested by considering the maximum value of DRHO+AVRHO over all cells in the flow domain. In theory it should be possible to reduce this figure to any arbitrarily small value by employing a sufficiently large number of time steps, but in practice one must expect a limit of about 10^{-5} or 10^{-6}, after which round-off errors will hamper any further convergence.

Depending upon the printing specifications in the input data, the program will print values of density, velocity, energy, pressure and temperature, at a given monitor point at selected time-steps. These results may appear on the screen (governed by the input parameter NPRIN1) or on the file entitled 'history.dat' (governed by the input parameter NPRIN2). The idea of storing the iterative history is that the file may later be studied at leisure; the nature of the iterative paths may suggest means to accelerate the convergence. The author of this book has already carried out extensive investigations in this area: Lobo (1992).

Once convergence to a desired accuracy has been achieved, the program prints the final solution - that is the values of density, velocity, pressure and temperature - at all cells on the flowfield. This is printed in File No 8 ('result.dat'). For a quick depiction of the solution however, it would be more useful to plot Mach number contours. Accordingly the program prints the local Mach numbers of the converged solution on two streamlines, one along each wall or, if the user so wishes, one along a wall and the other midway between the two walls (eg if the cascade were symmetric). These Mach numbers, along with the appropriate grid coordinates are stored in File No 9 ('graphin.dat') from where they may be read by a graphics subroutine.

Finally the program prints the 'converged' values of density, velocity and pressure in File No 10 ('continue.dat') in the same format as File No 3 ('inguess.dat'). This may then be used as a basis for continuation runs if desired.

The various subroutines are small and essentially self-explanatory. With the aid of the comment statements and the formulae derived earlier in this chapter, a reader should have no difficulty in understanding the steps involved.

It is now time to write the complete Fortran program for the Euler solver.

4 The Fortran listing

```
*********************************************************************
*     A TWO DIMENSIONAL TIME-MARCHING EULER SOLVER
*********************************************************************
C
        PARAMETER (M=32, M1=33, N=64, N1=65)
        COMMON /BLK0/ IM, KM, KLE, KTE
        COMMON /BLK1/ X(0:M,0:N), Z(0:M,0:N)
        COMMON /BLK2/ S1X(0:M,N), S1Z(0:M,N), S3X(M,0:N), S3Z(M,0:N)
        COMMON /BLK3/ VOL(M,N)
        COMMON /BLK4/ RHO(0:M1,0:N1), RHOT(0:M1,0:N1), DRHO(M,N)
        COMMON /BLK5/ RVX(0:M1,0:N1), RVXT(0:M1,0:N1), DRVX(M,N)
        COMMON /BLK6/ RVZ(0:M1,0:N1), RVZT(0:M1,0:N1), DRVZ(M,N)
        COMMON /BLK7/ RE(0:M1,0:N1), RET(0:M1,0:N1), DRE(M,N)
        COMMON /BLK8/ P(0:M1,0:N1), RH(0:M1,0:N1)
        COMMON /BLK9/ VX(0:M1,0:N1), VZ(0:M1,0:N1)
        COMMON /BLK10/ E(0:M1,0:N1), H(0:M1,0:N1), T(0:M1,0:N1)
        COMMON /BLK11/ Q1(0:M,N), Q3(M,0:N)
        COMMON /BLK12/ C(M,N), DT(M,N)
        COMMON /ART1/ AX1(0:M,N), AX3(0:M,N), AZ1(M,0:N), AZ3(M,0:N)
        COMMON /ART2/ AVRHO(M,N), AVRVX(M,N), AVRVZ(M,N), AVRE(M,N)
        COMMON /BOUND1/ISUPER,ISUBAX,PT,TT,THIN(M),PB,VSWIRL,XMACH
        COMMON /BOUND2/ RCS(N), RCP(N)
        COMMON /CONST/ CP, CV, GAMMA, GASCON
        DIMENSION CF(4)
        REAL MIDCEL(N),MACH(N)
C----------------------------------------------------------------
C.....  OPEN INPUT AND OUTPUT FILES
C.....  INPUT GRID FILE
        OPEN (1, FILE = 'grid.dat')
C.....  INPUT OPERATING CONDITIONS FILE
        OPEN (2, FILE = 'opercon.dat')
C.....  INPUT INITIAL GUESS FILE
        OPEN (3, FILE = 'inguess.dat')
```

```
C.....  OUTPUT CELL-AREAS FILE
        OPEN (4, FILE = 'areas.dat')
C.....  OUTPUT ITERATIVE-HISTORY FILE
        OPEN (7, FILE = 'history.dat')
C.....  OUTPUT SOLUTION FILE
        OPEN (8, FILE = 'result.dat')
C.....  OUTPUT GRAPHICS-INPUT FILE
        OPEN (9, FILE = 'graphin.dat')
C.....  OUTPUT FILE FOR CONTINUATION RUNS
        OPEN (10, FILE = 'continue.dat')
C------------------------------------------------------------------------
C.....  READ IN GRID SPECIFICATIONS
C       KM : THE LAST POINT IN STREAMWISE DIRECTION
C       IM : THE LAST POINT IN PITCHWISE DIRECTION
C       KLE: THE POINT AT LEADING EDGE
C       KTE : THE POINT AT TRAILING EDGE
C       LEVEL: INDEX (0 OR 1) STATING WHETHER WALL IS FLAT OR NOT
        READ (1,300) IM,KM,KLE,KTE,LEVEL
        DO 10 K = 0,KM
        DO 10 I = 0,IM
        READ (1,310) X(I,K), Z(I,K)
   10   CONTINUE
C------------------------------------------------------------------------
C.....  READ IN OPERATING CONDITIONS
C       ISUPER : INDEX FOR WHOLLY SUPERSONIC FLOW
C       ISUBAX : INDEX FOR SUPERSONIC FLOW WITH
C               SUBSONIC AXIAL COMPONENT
        READ (2,320) ISUPER, ISUBAX
C------------------------------------------------------------------------
C.....  READ IN BOUNDARY CONDITIONS
C       PT : TOTAL PRESSURE AT INLET
C       TT : TOTAL TEMPERATURE AT INLET
C       THIN : THETA(INCIDENCE ANGLE) AT INLET
C       PB : BACK PRESSURE (FOR SUBSONIC FLOW)
C       VSWIRL : SWIRL VELOCITY
C           (FOR SUPERSONIC FLOW WITH SUBSONIC AXIAL COMPONENT)
C       XMACH : INLET MACH NUMBER (FOR SUPERSONIC FLOW)
        READ (2,330) PT,TT
        READ (2,332) (THIN(I), I=1,IM)
        IF (ISUPER.EQ.1) GO TO 20
        READ (2,330) PB
         IF (ISUBAX.EQ.1) READ (2,334) VSWIRL
        GO TO 25
   20   READ (2,336) XMACH
   25   CONTINUE
C------------------------------------------------------------------------
```

```
C--------------------------------------------------------------------
C.....  READ IN TIME-FACTOR (PARAMETER FOR VARYING THE TIME STEP)
c.....  Note: This must be ≤ √3 for a 3-stage scheme and 2√2 for a 4-stage scheme
        READ (2,340) TIMFAC
C--------------------------------------------------------------------
C.....  READ IN COEFFICIENTS FOR ARTIFICIAL VISCOSITY
        READ (2,350) A1, A3
C--------------------------------------------------------------------
C.....  READ IN RUNGE-KUTTA SPECIFICATIONS
C       MSTAGE : NUMBER OF STAGES IN RUNGE-KUTTA SCHEME
C       CF: COEFFICIENTS IN RUNGE-KUTTA SCHEME
        READ (2,360) MSTAGE
        READ (2,365) (CF(I), I=1,MSTAGE)
C--------------------------------------------------------------------
C.....  READ IN PRINTING CRITERIA
C.....  IPR & KPR ARE THE COEFFICIENTS OF THE MONITOR POINT
C.....  (WHERE RESULTS WIL BE PRINTED ON THE SCREEN
C.....  AT REGULAR INTERVALS).
C.....  NPRIN1 IS THE TIME-STEP-INTERVAL AT WHICH MONITOR-POINT
C.....  RESULTS WILL BE PRINTED ON THE SCREEN.
C.....  NPRIN2 IS THE TIME-STEP-INTERVAL AT WHICH RESULTS
C.....  WILL BE PRINTED ON A FILE FOR LATER EXAMINATION.
        READ (2,370) IPR, KPR, NPRIN1, NPRIN2
C--------------------------------------------------------------------
C.....  READ IN TERMINATION CRITERIA
C       LCOUNT : MAXIMUM NUMBER OF TIME STEPS
C       EPSLN  : CONVERGENCE CRITERION
        READ (2,380) LCOUNT, EPSLN
C--------------------------------------------------------------------
C.....  READ IN INDEX FOR INITIAL ITERATION
        READ (2,320) INIT
C.....  IF INIT = 0, READ FROM THE 'INITIAL GUESS' FILE
        IF (INIT.EQ.0) THEN
        DO 30 K = 1,KM
        DO 30 I = 1,IM
        READ (3,390) RHO(I,K), VX(I,K), VZ(I,K), P(I,K)
   30   CONTINUE
        ICOUNT = 0
        ELSE
C.....  OTHERWISE READ FROM THE 'CONTINUATION FILE'
        DO 35 K = 1,KM
        DO 35 I = 1,IM
        READ (10,390) RHO(I,K), VX(I,K), VZ(I,K), P(I,K)
   35   CONTINUE
C.....  READ ITERATION NUMBER FROM 'CONTINUATION FILE'
        READ (10,395) ICOUNT
        END IF
C--------------------------------------------------------------------
```

```
C-------------------------------------------------------------------------
C.....  CHECK WHETHER INPUT HAS BEEN READ IN CORRECTLY
        CALL CHECK (IM,KM,KLE,KTE,LEVEL,ISUPER,ISUBAX,PT,TT,THIN,PB,
     1  VSWIRL,XMACH,TIMFAC,A1,A3,MSTAGE,CF,IPR,KPR,LCOUNT,EPSLN)
C-------------------------------------------------------------------------
C.....  DEFINE FUNDAMENTAL CONSTANTS
C       CP : SPECIFIC HEAT OF CONSTANT PRESSURE
C       CV : SPECIFIC HEAT OF CONSTANT VOLUME
C       GAMMA : RATIO OF SPECIFIC HEATS
C       GASCON : GAS CONSTANT
        CP = 1005.0
        CV = 718.0
        GAMMA = 1.4
        GASCON = 287.0
C-------------------------------------------------------------------------
C.....  CALCULATE THE AREA AND VOLUME OF EACH CELL (see Topic 2)
        CALL GEOM
        DO 40 K = 0,KM
        DO 40 I = 0,IM
        IF (I.GT.0.AND.K.GT.0) THEN
        WRITE (4,400) K, I, S1X(I,K), S1Z(I,K), S3X(I,K), S3Z(I,K), VOL(I,K)
        ELSE IF (I.EQ.0.AND.K.NE.0) THEN
        WRITE (4,400) K, I, S1X(I,K), S1Z(I,K)
        ELSE IF (K.EQ.0.AND.I.NE.0) THEN
        WRITE (4,405) K, I, S3X(I,K), S3Z(I,K)
        END IF
   40   CONTINUE
        WRITE (6,45)
   45   FORMAT (' CELL AREAS AND VOLUMES CALCULATED')
C-------------------------------------------------------------------------
C.....  COMPUTE RADII OF CURVATURE AT GRID POINTS (see Topic 2)
C.....  ON THE TWO SURFACES
        IF (LEVEL.EQ.0) GO TO 52
        CALL RADIUS
        DO 50 K = KLE,KTE-1
   50   WRITE (6,410) K, RCS(K), RCP(K)
        WRITE (6,55)
   52   CONTINUE
   55   FORMAT(' RADII OF CURVATURE CALCULATED')
C-------------------------------------------------------------------------
C.....  INITIALIZE TEMPERATURE,ENTHALPY AND ENERGY
        DO 60 K = 1,KM
        DO 60 I = 1,IM
        T(I,K) = P(I,K)/RHO(I,K)/GASCON
        HTV = 0.5*(VX(I,K)**2+VZ(I,K)**2)
        H(I,K) = CP*T(I,K) + HTV
        E(I,K) = CV*T(I,K) + HTV
   60   CONTINUE
```

115

```
C----------------------------------------------------------------------
C.....  IMPOSE 'FIXED BOUNDARY CONDITIONS', THAT IS, BOUNDARY
C.....  CONDITIONS WHICH WILL NOT ALTER AT EACH TIME STEP.
C.....  THE ONLY ONE IS BACK PRESSURE FOR SUBSONIC FLOW
        IF (ISUPER.EQ.1) GO TO 75
        DO 70 I = 1,IM
   70   P(I,KM+1) = PB
   75   CONTINUE
C----------------------------------------------------------------------
C.....  IMPOSE OTHER BOUNDARY CONDITIONS
        CALL INLET
        CALL PERIOD
        IF (LEVEL.EQ.0) CALL FLATWALL
        IF (LEVEL.NE.0) CALL CURWALL
        CALL OUTLET
C----------------------------------------------------------------------
C.....  INITIALIZE 'RUNNING VARIABLES'
        DO 80 K = 0,KM+1
        DO 80 I = 0,IM+1
        RVX(I,K) = RHO(I,K)*VX(I,K)
        RVZ(I,K) = RHO(I,K)*VZ(I,K)
        RH(I,K)  = RHO(I,K)*H(I,K)
        RE(I,K)  = RHO(I,K)*E(I,K)
   80   CONTINUE
C*********************************************************************
C.....  MARCHING PROCESS BEGINS
  100   ICOUNT = ICOUNT + 1
C----------------------------------------------------------------------
C.....  PLACE RUNNING VARIABLES IN TEMPORARY STORAGE
        DO 110 K = 0,KM+1
        DO 110 I = 0,IM+1
        RHOT(I,K) = RHO(I,K)
        RVXT(I,K) = RVX(I,K)
        RVZT(I,K) = RVZ(I,K)
        RET(I,K) = RE(I,K)
  110   CONTINUE
C----------------------------------------------------------------------
C.....  CALCULATE THE FLUX THROUGH EACH CELL
        CALL FLUX (IM,KM)
C----------------------------------------------------------------------
C.....  DETERMINE LOCAL SOUND SPEED
c.....  See Equation (3.34), page 57
        DO 120 K = 1,KM
        DO 120 I = 1,IM
        C(I,K) = SQRT(GAMMA*GASCON*T(I,K))
  120   CONTINUE
C----------------------------------------------------------------------
```

```
C-----------------------------------------------------------------------
C.....  DETERMINE LOCAL TIME STEP
c .....  Note:  This must be done in accordance with the two-dimensional CFL criterion
c .....         as explained in Topic 5.   TIMFAC must be ≤ √3 for a 3-stage scheme
c .....         and ≤ 2√2 for a 4-stage scheme as explained in Topic 8.
        DO 130 K = 1,KM
        DO 130 I = 1,IM
        SX1 = 0.5*(S1X(I-1,K)+S1X(I,K))
        SZ1 = 0.5*(S1Z(I-1,K)+S1Z(I,K))
        SX3 = 0.5*(S3X(I,K-1)+S3X(I,K))
        SZ3 = 0.5*(S3Z(I,K-1)+S3Z(I,K))
        F1 = ABS(VX(I,K)*SX1+VZ(I,K)*SZ1)
        F3 = ABS(VX(I,K)*SX3+VZ(I,K)*SZ3)
        CH1 = F1 + C(I,K)*SQRT(SX1*SX1+SZ1*SZ1)
        CH3 = F3 + C(I,K)*SQRT(SX3*SX3+SZ3*SZ3)
        DT(I,K) = TIMFAC * VOL(I,K)/(CH1+CH3)
  130   CONTINUE
C-----------------------------------------------------------------------
C.....  COMPUTE ARTIFICIAL VISCOSITY TERMS (see Topic 7)
        CALL ARTCF (A1,A3)
        CALL ARTVIS (RHO,AVRHO)
        CALL ARTVIS (RVX,AVRVX)
        CALL ARTVIS (RVZ,AVRVZ)
        CALL ARTVIS (RE,AVRE)
C-----------------------------------------------------------------------
C.....  THE NUMERICAL SOLVER: M-STAGE TIME-STEPPING SCHEME
c .....  (see Topic 8)
        IF (MSTAGE.EQ.1) GO TO 140
C
C.....  THE FIRST STAGE
        CALL DELRHO
        CALL DELRVX
        CALL DELRVZ
        CALL DELRE
        CALL CORRECTION (CF(1))
        CALL FLUX
        IF (MSTAGE.EQ.2) GO TO 140
C
C.....  THE SECOND STAGE
        CALL DELRHO
        CALL DELRVX
        CALL DELRVZ
        CALL DELRE
        CALL CORRECTION (CF(2))
        CALL FLUX
        IF (MSTAGE.EQ.3) GO TO 140
C-----------------------------------------------------------------------
```

```
C--------------------------------------------------------------------------------
C.....   THE THIRD STAGE
         CALL DELRHO
         CALL DELRVX
         CALL DELRVZ
         CALL DELRE
         CALL CORRECTION (CF(3))
         CALL FLUX
  140  CONTINUE
C
C.....   THE LAST STAGE
         CALL DELRHO
         CALL DELRVX
         CALL DELRVZ
         CALL DELRE
         CALL CORRECTION (CF(MSTAGE))
C--------------------------------------------------------------------------------
C.....   RE-SET DENSITY
         DO 150 K = 1,KM
         DO 150 I = 1,IM
  150  RHO(I,K) = RHOT(I,K)
C--------------------------------------------------------------------------------
C.....   IMPOSE THE INLET, PERIODIC, WALL AND OUTLET
C.....   BOUNDARY CONDITIONS (see Topic 9)
         CALL INLET
         CALL PERIOD
         IF (LEVEL.EQ.0) CALL FLATWALL
         IF (LEVEL.NE.0) CALL CURWALL
         CALL OUTLET
C--------------------------------------------------------------------------------
C.....   RE-SET RUNNING VARIABLES
         DO 160 K = 0,KM+1
         DO 160 I = 0,IM+1
         RVX(I,K) = RHO(I,K)*VX(I,K)
         RVZ(I,K) = RHO(I,K)*VZ(I,K)
         RH(I,K)  = RHO(I,K)*H(I,K)
         RE(I,K)  = RHO(I,K)*E(I,K)
  160  CONTINUE
C--------------------------------------------------------------------------------
```

```
C----------------------------------------------------------------------------------------
C.....   CHECK FOR CONVERGENCE
        RESDUE = 0.0
        SUM = 0.0
        DO 170 I = 1,IM
        DO 170 K = 1,KM
        ABDR = ABS(DRHO(I,K)+AVRHO(I,K))
        SUM = SUM + ABDR
        IF (ABDR.LE.RESDUE) GO TO 170
        IRES = I
        KRES = K
        RESDUE = ABDR
  170   CONTINUE
        IF (MOD(ICOUNT,NPRIN1).EQ.0)
     1  WRITE (6,420) ICOUNT,IRES,KRES,RESDUE,
     2  RHO(IPR,KPR),VX(IPR,KPR),VZ(IPR,KPR),E(IPR,KPR),
     3  P(IPR,KPR),T(IPR,KPR)
        IF (MOD(ICOUNT,NPRIN2).EQ.0) THEN
        WRITE (7,420) ICOUNT,IRES,KRES,AVDR,
     3  RHO(IPR,KPR),VX(IPR,KPR),VZ(IPR,KPR),E(IPR,KPR),
     4  P(IPR,KPR),T(IPR,KPR)
        END IF
C----------------------------------------------------------------------------------------
C.....   PRINT FINAL SOLUTION
        WRITE (6,500)
        DO 180 K = 0,KM+1
        DO 180 I = 0,IM+1
        WRITE (8,430) I,K,RHO(I,K),VX(I,K),VZ(I,K),P(I,K),T(I,K)
  180   CONTINUE
C.....   PRINT SELECTED RESULTS FOR GRAPHICS PLOTS
        IMB2 = IM/2
        DO 200 I = 1,IMB2,IMB2-1
        DO 190 K = 1,KM
        MIDCEL(K) = 0.5*(Z(I,K-1)+Z(I,K))
        VTOT = SQRT(VX(I,K)**2+VZ(I,K)**2)
  190   MACH(K) = VTOT/C(I,K)
        WRITE (9,*) (MIDCEL(K),K=1,KM)
  200   WRITE (9,*) (MACH(K),K=1,KM)
C.....   PRINT SOLUTION IN AN INPUT-FORM FOR POSIBLE
C.....   CONTINUATION RUNS
        DO 210 K = 1,KM
        DO 210 I = 1,IM
        WRITE (10,390) RHO(I,K),VX(I,K),VZ(I,K),P(I,K)
  210   CONTINUE
        WRITE (10,395) ICOUNT
C----------------------------------------------------------------------------------------
```

119

```
C------------------------------------------------------------------------
C.....  CLOSE ALL FILES AND STOP
        CLOSE(1)
        CLOSE(2)
        CLOSE(3)
        CLOSE(4)
        CLOSE(7)
        CLOSE(8)
        CLOSE(9)
        CLOSE(10)
        STOP
C------------------------------------------------------------------------
C.....  FORMATS FOR READ STATEMENTS
  300   FORMAT (5I5)
  310   FORMAT (2F12.6)
  320   FORMAT (2I2)
  330   FORMAT (F9.1,F7.2)
  332   FORMAT (12F6.3)
  334   FORMAT (F9.2)
  336   FORMAT (F7.4)
  340   FORMAT (F4.2)
  350   FORMAT (2F5.3)
  360   FORMAT (I1)
  365   FORMAT (4F5.2)
  370   FORMAT (4I3)
  380   FORMAT (I4,F9.6)
  390   FORMAT (F12.6,2F12.4,F12.2)
  395   FORMAT (I5)
C------------------------------------------------------------------------
C.....  FORMATS FOR WRITE STATEMENTS
  400   FORMAT (2I5,8F8.4)
  405   FORMAT (2I5,16X,2F8.4)
  410   FORMAT (I5,' RCS = ',F6.4,' : RCP = ',F6.4)
  420   FORMAT (I4,':',2I3,F9.6,' :',F9.6,2F9.4,2F9.1,F9.4)
  430   FORMAT (2I5,F10.6,2F10.4,F10.1,F10.4,F10.6)
  440   FORMAT (I5,2(F9.4,2F9.2,F9.0,' : '))
  450   FORMAT (I5,4F8.4,2(' : ',F9.6))
  460   FORMAT (F9.4,2F9.2,F9.0)
C------------------------------------------------------------------------
  500   FORMAT (' THE ITERATES HAVE CONVERGED')
        END
```

```
***********************************************************************
*                                                                     *
      SUBROUTINE CHECK
    1 (IM,KM,KLE,KTE,LEVEL,ISUPER,ISUBAX,PT,TT,THIN,PB,
    2 VSWIRL,XMACH,TIMFAC,A1,A3,MSTAGE,CF,IPR,KPR,LCOUNT,EPSLN)
*                                                                     *
***********************************************************************
C
c.....   This subroutine simply prints the input data on the screen so that the user can
c.....   confirm that it has been fed in correctly.
C
      PARAMETER (M=32)
      DIMENSION THIN(M),CF(4)
      WRITE (6,10) IM
      WRITE (6,20) KM
      WRITE (6,30) KLE
      WRITE (6,40) KTE
      WRITE (6,50) LEVEL
      WRITE (6,60) ISUPER
      WRITE (6,70) ISUBAX
      WRITE (6,80) PT
      WRITE (6,90) TT
      WRITE (6,100) (THIN(I),I=1,IM)
      IF (ISUPER.EQ.0) WRITE (6,110) PB
      IF (ISUBAX.EQ.1) WRITE (6,120) VSWIRL
      IF (ISUPER.EQ.1) WRITE (6,130) XMACH
      WRITE (6,140) TIMFAC
      WRITE (6,150) A1
      WRITE (6,160) A3
      WRITE (6,170) MSTAGE
      WRITE (6,180) (CF(I),I=1,MSTAGE)
      WRITE (6,190) IPR
      WRITE (6,200) KPR
      WRITE (6,210) LCOUNT
      WRITE (6,220) EPSLN
      RETURN
   10 FORMAT ('     IM = ',I5)
   20 FORMAT ('     KM = ',I5)
   30 FORMAT ('    KLE = ',I5)
   40 FORMAT ('    KTE = ',I5)
   50 FORMAT ('  LEVEL = ',I5)
   60 FORMAT (' ISUPER = ',I5)
   70 FORMAT (' ISUBAX = ',I5)
   80 FORMAT ('     PT = ',F8.1)
   90 FORMAT ('     TT = ',F7.2)
  100 FORMAT ('   THIN = ',11F6.3)
```

```
110  FORMAT ('     PB = ',F9.2)
120  FORMAT (' VSWIRL = ',F8.3)
130  FORMAT ('  XMACH = ',F7.4)
140  FORMAT (' TIME FACTOR = ',F4.2)
150  FORMAT ('     A1 = ',F5.3)
160  FORMAT ('     A3 = ',F5.3)
170  FORMAT (' MSTAGE = ',I3)
180  FORMAT ('  CF(I) = ',4F5.2)
190  FORMAT ('    IPR = ',I3)
200  FORMAT ('    KPR = ',I3)
210  FORMAT (' LCOUNT = ',I5)
220  FORMAT ('  EPSLN = ',F9.6)
     END
C
************************************************************************
*                                                                    *
       SUBROUTINE GEOM
*                                                                    *
************************************************************************
C
c.....  The formulae used in this subroutine are derived in Topic 2:
c.....  Equations (3.1) for the projective lengths and Equation (3.5b) for the area.
C
       PARAMETER (M=32,N=64)
       COMMON /BLK0/ IM,KM,KLE,KTE
       COMMON /BLK1/ X(0:M,0:N),Z(0:M,0:N)
       COMMON /BLK2/ S1X(0:M,N),S1Z(0:M,N),S3X(M,0:N),S3Z(M,0:N)
       COMMON /BLK3/ VOL(M,N)
       DO 10 K = 1,KM
       DO 10 I = 0,IM
       S1X(I,K) = Z(I,K)-Z(I,K-1)
       S1Z(I,K) = X(I,K-1)-X(I,K)
  10   CONTINUE
       DO 20 K = 0,KM
       DO 20 I = 1,IM
       S3X(I,K) = Z(I-1,K)-Z(I,K)
       S3Z(I,K) = X(I,K)-X(I-1,K)
  20   CONTINUE
       DO 30 K = 1,KM
       DO 30 I = 1,IM
       X1 = X(I,K)-X(I-1,K-1)
       Z1 = Z(I,K)-Z(I-1,K-1)
       X2 = X(I-1,K)-X(I,K-1)
       Z2 = Z(I-1,K)-Z(I,K-1)
       VOL(I,K) = 0.5*(X1*Z2-Z1*X2)
  30   CONTINUE
       END
************************************************************************
```

```
********************************************************************
*                                                                  *
        SUBROUTINE RADIUS
*                                                                  *
********************************************************************
C
c.....  The formula used in this subroutine is derived in Topic 2: Equation (3.13).
c.....  The subroutine actually computes the inverse radius of curvature
c.....  as explained on page 48.  In fact SUBPROGRAM 3.4 (page 48) is used
c.....  at all points on the suction and pressure surfaces.
C
        PARAMETER (M=32,N=64)
        COMMON /BLK0/ IM,KM,KLE,KTE
        COMMON /BLK1/ X(0:M,0:N),Z(0:M,0:N)
        COMMON /BOUND2/ RCS(N),RCP(N)
C---------------------------------------------------------------------
C.....  THE SUCTION SURFACE
        DO 10 K = KLE,KTE-1
        A = SQRT((X(0,K)-X(0,K-1))**2+(Z(0,K)-Z(0,K-1))**2)
        B = SQRT((X(0,K+1)-X(0,K))**2+(Z(0,K+1)-Z(0,K))**2)
        C = SQRT((X(0,K+1)-X(0,K-1))**2+(Z(0,K+1)-Z(0,K-1))**2)
        ABC = A + B + C
        AB  = A + B - C
        BC  = B + C - A
        CA  = C + A - B
   10   RCS(K) = SQRT(AB*BC*CA*ABC)/(A*B*C)
C---------------------------------------------------------------------
C.....  THE PRESSURE SURFACE
        DO 20 K = KLE,KTE-1
        A = SQRT((X(IM,K)-X(IM,K-1))**2+(Z(IM,K)-Z(IM,K-1))**2)
        B = SQRT((X(IM,K+1)-X(IM,K))**2+(Z(IM,K+1)-Z(IM,K))**2)
        C = SQRT((X(IM,K+1)-X(IM,K-1))**2+(Z(IM,K+1)-Z(IM,K-1))**2)
        ABC = A + B + C
        AB  = A + B - C
        BC  = B + C - A
        CA  = C + A - B
   20   RCP(K) = SQRT(AB*BC*CA*ABC)/(A*B*C)
        END
********************************************************************
```

```
***********************************************************************
*                                                                     *
          SUBROUTINE DELRHO
*                                                                     *
***********************************************************************
C
c.....    This subroutine computes the increment Δρ in accordance with
c.....    Equation (3.69a) (page 72)
C
          PARAMETER (M=32,M1=33,N=64,N1=65)
          COMMON /BLK0/ IM,KM,KLE,KTE
          COMMON /BLK3/ VOL(M,N)
          COMMON /BLK4/ RHO(0:M1,0:N1),RHOT(0:M1,0:N1),DRHO(M,N)
          COMMON /BLK11/ Q1(0:M,N),Q3(M,0:N)
          COMMON /BLK12/ C(M,N),DT(M,N)
          DO 30 K = 1,KM
          DO 30 I = 1,IM
          FL1IN = 0.0
          IF (I.EQ.1.AND.K.GE.KLE.AND.K.LE.KTE) GO TO 10
          FL1IN = 0.5*(RHOT(I-1,K)+RHOT(I,K))*Q1(I-1,K)
   10     CONTINUE
          FL3IN = 0.5*(RHOT(I,K-1)+RHOT(I,K))*Q3(I,K-1)
          FL1OUT = 0.0
          IF (I.EQ.IM.AND.K.GE.KLE.AND.K.LE.KTE) GO TO 20
          FL1OUT = 0.5*(RHOT(I,K)+RHOT(I+1,K))*Q1(I,K)
   20     CONTINUE
          FL3OUT = 0.5*(RHOT(I,K)+RHOT(I,K+1))*Q3(I,K)
          DRHO(I,K) = (FL1IN+FL3IN-FL1OUT-FL3OUT) * DT(I,K)/VOL(I,K)
   30     CONTINUE
          RETURN
          END
***********************************************************************
```

```
***********************************************************************
*                                                                     *
        SUBROUTINE DELRVX
*                                                                     *
***********************************************************************
C
c.....   This subroutine computes the increment Δρu in accordance with
c.....   Equation (3.69b) (page 72)
C
        PARAMETER (M=32,M1=33,N=64,N1=65)
        COMMON /BLK0/ IM,KM,KLE,KTE
        COMMON /BLK2/ S1X(0:M,N),S1Z(0:M,N),S3X(M,0:N),S3Z(M,0:N)
        COMMON /BLK3/ VOL(M,N)
        COMMON /BLK5/ RVX(0:M1,0:N1),RVXT(0:M1,0:N1),DRVX(M,N)
        COMMON /BLK8/ P(0:M1,0:N1),RH(0:M1,0:N1)
        COMMON /BLK11/ Q1(0:M,N),Q3(M,0:N)
        COMMON /BLK12/ C(M,N),DT(M,N)
        DO 30 K = 1,KM
        DO 30 I = 1,IM
        FL1IN = 0.5*(P(I-1,K)+P(I,K))*S1X(I-1,K)
        IF (I.EQ.1.AND.K.GE.KLE.AND.K.LE.KTE) GO TO 10
        FL1IN = FL1IN + 0.5*(RVXT(I-1,K)+RVXT(I,K))*Q1(I-1,K)
  10    CONTINUE
        FL3IN = 0.5*((RVXT(I,K-1)+RVXT(I,K))*Q3(I,K-1)
     1             +(P(I,K-1)+P(I,K))*S3X(I,K-1))
        FL1OUT = 0.5*(P(I,K)+P(I+1,K))*S1X(I,K)
        IF (I.EQ.IM.AND.K.GE.KLE.AND.K.LE.KTE) GO TO 20
        FL1OUT = FL1OUT + 0.5*(RVXT(I,K)+RVXT(I+1,K))*Q1(I,K)
  20    CONTINUE
        FL3OUT = 0.5*((RVXT(I,K)+RVXT(I,K+1))*Q3(I,K)
     2             +(P(I,K)+P(I,K+1))*S3X(I,K))
        DRVX(I,K) = (FL1IN+FL3IN-FL1OUT-FL3OUT) * DT(I,K)/VOL(I,K)
  30    CONTINUE
        RETURN
        END
***********************************************************************
```

```
*********************************************************************
*                                                                   *
        SUBROUTINE DELRVZ
*                                                                   *
*********************************************************************
C
c.....  This subroutine computes the increment Δρv in accordance with
c.....  Equation (3.69c) (page 72)
C
        PARAMETER (M=32,M1=33,N=64,N1=65)
        COMMON /BLK0/ IM,KM,KLE,KTE
        COMMON /BLK2/ S1X(0:M,N),S1Z(0:M,N),S3X(M,0:N),S3Z(M,0:N)
        COMMON /BLK3/ VOL(M,N)
        COMMON /BLK6/ RVZ(0:M1,0:N1),RVZT(0:M1,0:N1),DRVZ(M,N)
        COMMON /BLK8/ P(0:M1,0:N1),RH(0:M1,0:N1)
        COMMON /BLK11/ Q1(0:M,N),Q3(M,0:N)
        COMMON /BLK12/ C(M,N),DT(M,N)
        DO 30 K = 1,KM
        DO 30 I = 1,IM
        FL1IN = 0.5*(P(I-1,K)+P(I,K))*S1Z(I-1,K)
        IF (I.EQ.1.AND.K.GE.KLE.AND.K.LE.KTE) GO TO 10
        FL1IN = FL1IN + 0.5*(RVZT(I-1,K)+RVZT(I,K))*Q1(I-1,K)
   10   CONTINUE
        FL3IN = 0.5*((RVZT(I,K-1)+RVZT(I,K))*Q3(I,K-1)
      1          +(P(I,K-1)+P(I,K))*S3Z(I,K-1))
         FL1OUT = 0.5*(P(I,K)+P(I+1,K))*S1Z(I,K)
        IF (I.EQ.IM.AND.K.GE.KLE.AND.K.LE.KTE) GO TO 20
        FL1OUT = FL1OUT + 0.5*(RVZT(I,K)+RVZT(I+1,K))*Q1(I,K)
   20   CONTINUE
        FL3OUT = 0.5*((RVZT(I,K)+RVZT(I,K+1))*Q3(I,K)
      2          +(P(I,K)+P(I,K+1))*S3Z(I,K))
        DRVZ(I,K) = (FL1IN+FL3IN-FL1OUT-FL3OUT) * DT(I,K)/VOL(I,K)
   30   CONTINUE
        RETURN
        END
*********************************************************************
```

```
************************************************************************
*                                                                      *
         SUBROUTINE DELRE
*                                                                      *
************************************************************************
C
c.....   This subroutine computes the increment Δρe in accordance with
c.....   Equation (3.69d) (page 72)
C
         PARAMETER (M=32,M1=33,N=64,N1=65)
         COMMON /BLK0/ IM,KM,KLE,KTE
         COMMON /BLK3/ VOL(M,N)
         COMMON /BLK7/ RE(0:M1,0:N1),RET(0:M1,0:N1),DRE(M,N)
         COMMON /BLK8/ P(0:M1,0:N1),RH(0:M1,0:N1)
         COMMON /BLK11/ Q1(0:M,N),Q3(M,0:N)
         COMMON /BLK12/ C(M,N),DT(M,N)
         DO 30 K = 1,KM
         DO 30 I = 1,IM
         FL1IN = 0.0
         IF (I.EQ.1.AND.K.GE.KLE.AND.K.LE.KTE) GO TO 10
         FL1IN = 0.5*(RH(I-1,K)+RH(I,K))*Q1(I-1,K)
   10    CONTINUE
         FL3IN = 0.5*(RH(I,K-1)+RH(I,K))*Q3(I,K-1)
         FL1OUT = 0.0
         IF (I.EQ.IM.AND.K.GE.KLE.AND.K.LE.KTE) GO TO 20
         FL1OUT = 0.5*(RH(I,K)+RH(I+1,K))*Q1(I,K)
   20    CONTINUE
         FL3OUT = 0.5*(RH(I,K)+RH(I,K+1))*Q3(I,K)
         DRE(I,K) = (FL1IN+FL3IN-FL1OUT-FL3OUT) * DT(I,K)/VOL(I,K)
   30    CONTINUE
         RETURN
         END
************************************************************************
```

127

```
**********************************************************************
*                                                                    *
      SUBROUTINE ARTCF (A1,A3)
*                                                                    *
**********************************************************************
C
c.....  This subroutine computes the artificial coefficients a_{x1} and a_{x3} in accordance
c.....  with Equations (3.73) & (3.75) (page 78). The artificial coefficients a_{z1} and a_{z3}
c.....  are obtained in a similar manner .
C
      PARAMETER (M=32,M1=33,N=64,N1=65)
      COMMON /BLK0/ IM,KM,KLE,KTE
      COMMON /BLK8/ P(0:M1,0:N1),RH(0:M1,0:N1)
      COMMON /ART1/ AX1(0:M,N),AX3(0:M,N),AZ1(M,0:N),AZ3(M,0:N)
C----------------------------------------------------------------------
C.....  COMPUTE THE COEFFICIENTS AX1 & AX3 AT HORIZONTAL FACES
      DO 20 K = 1,KM
      PXX = ABS(P(2,K)-2*P(1,K)+P(0,K))/(P(2,K)+2*P(1,K)+P(0,K))
      AX1(0,K) = A1*PXX
      DO 10 I = 1,IM-1
      PXXOLD = PXX
      PXX = ABS(P(I+1,K)-2*P(I,K)+P(I-1,K))/(P(I+1,K)+2*P(I,K)+P(I-1,K))
   10 AX1(I,K) = A1*AMAX1(PXXOLD,PXX)
      AX1(IM,K) = A1*PXX
      DO 20 I = 0,IM
   20 AX3(I,K) = AMAX1(A3-AX1(I,K),0.0)
C----------------------------------------------------------------------
C.....  COMPUTE THE COEFFICIENTS AZ1 & AZ3 AT VERTICAL FACES
      DO 40 I = 1,IM
      PZZ = ABS(P(I,2)-2*P(I,1)+P(I,0))/(P(I,2)+2*P(I,1)+P(I,0))
      AZ1(I,0) = A1*PZZ
      DO 30 K = 1,KM-1
      PZZOLD = PZZ
      PZZ = ABS(P(I,K+1)-2*P(I,K)+P(I,K-1))/(P(I,K+1)+2*P(I,K)+P(I,K-1))
   30 AZ1(I,K) = A1*AMAX1(PZZOLD,PZZ)
      AZ1(I,KM) = A1*PZZ
      DO 40 K = 0,KM
   40 AZ3(I,K) = AMAX1(A3-AZ1(I,K),0.0)
      RETURN
      END
**********************************************************************
```

```
********************************************************************
*                                                                *
       SUBROUTINE ARTVIS (F,AVF)
*                                                                *
********************************************************************
C
c.....  This subroutine computes the dissipative contribution Af in accordance with
c.....  Equation (3.72) (page 77). Contributions in the x and z directions are added,
c.....  and the subroutine is called for each of the four fundamental variables.
C
       PARAMETER (M=32,M1=33,N=64,N1=65)
       COMMON /BLK0/ IM,KM,KLE,KTE
       COMMON /ART1/ AX1(0:M,N),AX3(0:M,N),AZ1(M,0:N),AZ3(M,0:N)
       DIMENSION F(0:M1,0:N1),AVF(M,N)
C--------------------------------------------------------------------
C.....  COMPUTE THE CONTRIBUTION FROM THE X-DIRECTION
       DO 20 K = 1,KM
       FX1 = F(1,K)-F(0,K)
       FX3 = F(3,K)-3*F(2,K)+3*F(1,K)-F(0,K)
       FX = AX1(0,K)*FX1-AX3(0,K)*FX3
       DO 20 I = 1,IM
       FXOLD = FX
       FX1 = F(I+1,K)-F(I,K)
       IF (I.EQ.IM) GO TO 10
       FX3 = F(I+2,K)-3*F(I+1,K)+3*F(I,K)-F(I-1,K)
   10  FX = AX1(I,K)*FX1-AX3(I,K)*FX3
   20  AVF(I,K) = FX-FXOLD
C--------------------------------------------------------------------
C.....  COMPUTE THE CONTRIBUTION FROM THE Z-DIRECTION
       DO 40 I = 1,IM
       FZ1 = F(I,1)-F(I,0)
       FZ3 = F(I,3)-3*F(I,2)+3*F(I,1)-F(I,0)
       FZ = AZ1(I,0)*FZ1-AZ3(I,0)*FZ3
       DO 40 K = 1,KM
       FZOLD = FZ
       FZ1 = F(I,K+1)-F(I,K)
       IF (K.EQ.KM) GO TO 30
       FZ3 = F(I,K+2)-3*F(I,K+1)+3*F(I,K)-F(I,K-1)
   30  FZ = AZ1(I,K)*FZ1-AZ3(I,K)*FZ3
   40  AVF(I,K) = AVF(I,K) + FZ-FZOLD
       RETURN
       END
********************************************************************
```

```
*************************************************************************
*                                                                       *
         SUBROUTINE CORRECTION (CF)
*                                                                       *
*************************************************************************
C
         PARAMETER (M=32,M1=33,N=64,N1=65)
         COMMON /BLK0/ IM,KM,KLE,KTE
         COMMON /BLK4/ RHO(0:M1,0:N1),RHOT(0:M1,0:N1),DRHO(M,N)
         COMMON /BLK5/ RVX(0:M1,0:N1),RVXT(0:M1,0:N1),DRVX(M,N)
         COMMON /BLK6/ RVZ(0:M1,0:N1),RVZT(0:M1,0:N1),DRVZ(M,N)
         COMMON /BLK7/ RE(0:M1,0:N1),RET(0:M1,0:N1),DRE(M,N)
         COMMON /BLK8/ P(0:M1,0:N1),RH(0:M1,0:N1)
         COMMON /BLK9/ VX(0:M1,0:N1),VZ(0:M1,0:N1)
         COMMON /BLK10/ E(0:M1,0:N1),H(0:M1,0:N1),T(0:M1,0:N1)
         COMMON /ART2/ AVRHO(M,N),AVRVX(M,N),AVRVZ(M,N),AVRE(M,N)
         COMMON /CONST/ CP,CV,GAMMA,GASCON
         DO 10 K = 1,KM
         DO 10 I = 1,IM
         RHOT(I,K) = RHO(I,K)+CF*(DRHO(I,K)+AVRHO(I,K))
         RVXT(I,K) = RVX(I,K)+CF*(DRVX(I,K)+AVRVX(I,K))
         RVZT(I,K) = RVZ(I,K)+CF*(DRVZ(I,K)+AVRVZ(I,K))
         RET(I,K)  = RE(I,K)+CF*(DRE(I,K)+AVRE(I,K))
         VX(I,K) = RVXT(I,K)/RHOT(I,K)
         VZ(I,K) = RVZT(I,K)/RHOT(I,K)
         E(I,K)  = RET(I,K)/RHOT(I,K)
         HTV = 0.5*(VX(I,K)**2+VZ(I,K)**2)
         T(I,K) = (E(I,K)-HTV)/CV
         IF (T(I,K).LT.0.0) GO TO 20
         P(I,K) = RHOT(I,K)*GASCON*T(I,K)
         H(I,K) = CP*T(I,K) + HTV
         RH(I,K) = RHOT(I,K)*H(I,K)
   10    CONTINUE
         RETURN
   20    WRITE (6,30) I,K
         WRITE (6,40) DRHO(I,K),RHO(I,K)
         WRITE (6,50) DRVX(I,K),VX(I,K)
         WRITE (6,60) DRVZ(I,K),VZ(I,K)
         WRITE (6,70) DRE(I,K),E(I,K)
         STOP
   30    FORMAT (' THE TEMPERATURE IS NEGATIVE AT (',I2,',',I2,')')
   40    FORMAT (' DRHO = ',F9.6,' : RHO = ',F9.6)
   50    FORMAT (' DRVX = ',F9.4,' :  VX = ',F9.4)
   60    FORMAT (' DRVZ = ',F9.4,' :  VZ = ',F9.4)
   70    FORMAT (' DRE  = ',F9.1,' :   E = ',F9.1)
         END
*************************************************************************
```

```
************************************************************************
*                                                                     *
         SUBROUTINE FLUX
*                                                                     *
************************************************************************
C
c.....   This subroutine computes the mass fluxes through the cell faces.
c.....   The quantities Q₁ and Q₂ are defined on page 69.
C
         PARAMETER (M=32,M1=33,N=64,N1=65)
         COMMON /BLK0/ IM,KM,KLE,KTE
         COMMON /BLK2/ S1X(0:M,N),S1Z(0:M,N),S3X(M,0:N),S3Z(M,0:N)
         COMMON /BLK9/ VX(0:M1,0:N1),VZ(0:M1,0:N1)
         COMMON /BLK11/ Q1(0:M,N),Q3(M,0:N)
         DO 10 K = 1,KM
         DO 10 I = 0,IM
         IF (I.EQ.0.OR.I.EQ.IM) THEN
         IF (K.GE.KLE.AND.K.LE.KTE) GO TO 10
         END IF
         Q1(I,K) = 0.5*((VX(I,K)+VX(I+1,K))*S1X(I,K)
     1                 +(VZ(I,K)+VZ(I+1,K))*S1Z(I,K))
   10    CONTINUE
         DO 20 K = 0,KM
         DO 20 I = 1,IM
         Q3(I,K) = 0.5*((VX(I,K)+VX(I,K+1))*S3X(I,K)
     2                 +(VZ(I,K)+VZ(I,K+1))*S3Z(I,K))
   20    CONTINUE
         RETURN
         END
************************************************************************
```

```
******************************************************************
*                                                              *
        SUBROUTINE INLET
*                                                              *
******************************************************************
C
c.....  This subroutine applies the inlet boundary conditions in accordance with
c.....  Equations (3.32) (page 56), Equations (3.102) to (3.105) (page 93)
c.....  and Equations (2.8) & (2.9) (page 26).
C
        PARAMETER (M=32,M1=33,N=64,N1=65)
        COMMON /BLK0/ IM,KM,KLE,KTE
        COMMON /BLK4/ RHO(0:M1,0:N1),RHOT(0:M1,0:N1),DRHO(M,N)
        COMMON /BLK8/ P(0:M1,0:N1),RH(0:M1,0:N1)
        COMMON /BLK9/ VX(0:M1,0:N1),VZ(0:M1,0:N1)
        COMMON /BLK10/ E(0:M1,0:N1),H(0:M1,0:N1),T(0:M1,0:N1)
        COMMON /BOUND1/ISUPER,ISUBAX,PT,TT,THIN(M),PB,VSWIRL,XMACH
        COMMON /CONST/ CP,CV,GAMMA,GASCON
        DO 10 I = 1,IM
        IF (ISUPER.NE.1) THEN
        VTOT1 = SQRT(VX(I,1)**2+VZ(I,1)**2)
        VTOT2 = SQRT(VX(I,2)**2+VZ(I,2)**2)
        VTOT0 = 2*VTOT1-VTOT2
        T(I,0) = TT-0.5*VTOT1**2/CP
        C = SQRT(GAMMA*GASCON*T(I,0))
        XMACH = VTOT0/C
        ELSE
        T(I,0) = TT/(1.+.5*(GAMMA-1.)*XMACH**2)
        C = SQRT(GAMMA*GASCON*T(I,0))
        VTOT0 = XMACH*C
        END IF
        P(I,0) = PT/(1.+.5*(GAMMA-1.)*XMACH**2)**(GAMMA/(GAMMA-1.))
        RHO(I,0) = P(I,0)/(GASCON*T(I,0))
        IF (ISUBAX.NE.1) THEN
        VZ(I,0) = VTOT0*COS(THIN(I))
        VX(I,0) = VTOT0*SIN(THIN(I))
        ELSE
        VX(I,0) = VSWIRL
        VZ(I,0) = SQRT(VTOT0**2-VSWIRL**2)
        END IF
        HTV = 0.5*VTOT0**2
        H(I,0) = CP*T(I,0) + HTV
        E(I,0) = CV*T(I,0) + HTV
   10   CONTINUE
        RETURN
        END
******************************************************************
```

```
**********************************************************************
*                                                                    *
      SUBROUTINE PERIOD
*                                                                    *
**********************************************************************
C
c.....  This subroutine applies the periodic boundary conditions in accordance
c.....  with the explanation given in the first paragraph of page 94.
C
      PARAMETER (M=32,M1=33,N=64,N1=65)
      COMMON /BLK0/ IM,KM,KLE,KTE
      COMMON /BLK4/ RHO(0:M1,0:N1),RHOT(0:M1,0:N1),DRHO(M,N)
      COMMON /BLK8/ P(0:M1,0:N1),RH(0:M1,0:N1)
      COMMON /BLK9/ VX(0:M1,0:N1),VZ(0:M1,0:N1)
      COMMON /BLK10/ E(0:M1,0:N1),H(0:M1,0:N1),T(0:M1,0:N1)
      DO 10 K = 1,KM
      IF ((K.GE.KLE).AND.(K.LE.KTE)) GO TO 10
      RHO(0,K) = RHO(IM,K)
      VX(0,K)  = VX(IM,K)
      VZ(0,K)  = VZ(IM,K)
      E(0,K)   = E(IM,K)
      P(0,K)   = P(IM,K)
      H(0,K)   = H(IM,K)
      T(0,K)   = T(IM,K)
      RHO(IM+1,K) = RHO(1,K)
      VX(IM+1,K)  = VX(1,K)
      VZ(IM+1,K)  = VZ(1,K)
      E(IM+1,K)   = E(1,K)
      P(IM+1,K)   = P(1,K)
      H(IM+1,K)   = H(1,K)
      T(IM+1,K)   = T(1,K)
   10 CONTINUE
      RETURN
      END
**********************************************************************
```

```
*********************************************************************
*                                                                   *
        SUBROUTINE FLATWALL
*                                                                   *
*********************************************************************
C
c.....  This subroutine applies the wall boundary condition for a flat surface.
c.....  This will be characterized by a normal pressure gradient equal to zero
C
        PARAMETER (M=32,M1=33,N=64,N1=65)
        COMMON /BLK0/ IM,KM,KLE,KTE
        COMMON /BLK1/ X(0:M,0:N),Z(0:M,0:N)
        COMMON /BLK4/ RHO(0:M1,0:N1),RHOT(0:M1,0:N1),DRHO(M,N)
        COMMON /BLK8/ P(0:M1,0:N1),RH(0:M1,0:N1)
        COMMON /BLK9/ VX(0:M1,0:N1),VZ(0:M1,0:N1)
        COMMON /BLK10/ E(0:M1,0:N1),H(0:M1,0:N1),T(0:M1,0:N1)
C
C---------------------------------------------------------------------
C.....  THE SUCTION SURFACE
        DO 30 K = KLE,KTE
        SL = (X(0,K)-X(0,K-1))/(Z(0,K)-Z(0,K-1))
        SQL = SQRT(1.+SL**2)
        XS = 0.5*(X(0,K-1)+X(0,K))
        ZS = 0.5*(Z(0,K-1)+Z(0,K))
        XM = 0.5*(X(1,K-1)+X(1,K))
        ZM = 0.5*(Z(1,K-1)+Z(1,K))
        SLG = -(ZM-ZS)/(XM-XS)
        VTOTG = SQRT(VX(1,K)**2+VZ(1,K)**2)
        IF (SL.EQ.SLG) GO TO 10
        XG = 0.25*(X(0,K-1)+X(0,K)+X(1,K-1)+X(1,K))
        ZG = 0.25*(Z(0,K-1)+Z(0,K)+Z(1,K-1)+Z(1,K))
        XB = 2*XS-XG
        ZB = 2*ZS-ZG
        IF (SL.GT.SLG) KN = K-1
        IF (SL.LT.SLG) KN = K+1
        XN = 0.25*(X(0,KN-1)+X(0,KN)+X(1,KN-1)+X(1,KN))
        ZN = 0.25*(Z(0,KN-1)+Z(0,KN)+Z(1,KN-1)+Z(1,KN))
        VTOTN = SQRT(VX(1,KN)**2+VZ(1,KN)**2)
        RATIO = ((ZB-ZG)+SL*(XB-XG))/((ZN-ZG)+SL*(XN-XG))
        P(0,K) = P(1,K) + RATIO*(P(1,KN)-P(1,K))
        RHO(0,K) = RHO(1,K) + RATIO*(RHO(1,KN)-RHO(1,K))
        VTOT = VTOTG + RATIO*(VTOTN-VTOTG)
        E(0,K) = E(1,K) + RATIO*(E(1,KN)-E(1,K))
        GO TO 20
```

```
 10   P(0,K) = P(1,K)
      RHO(0,K) = RHO(1,K)
      VTOT = VTOTG
      E(0,K) = E(1,K)
 20   VX(0,K) = VTOT*SL/SQL
 30   VZ(0,K) = VTOT/SQL
C
C-----------------------------------------------------------------------------
C.....   THE PRESSURE SURFACE
      DO 60 K = KLE,KTE
      SL = (X(IM,K)-X(IM,K-1))/(Z(IM,K)-Z(IM,K-1))
      SQL = SQRT(1.+SL**2)
      XS = 0.5*(X(IM,K-1)+X(IM,K))
      ZS = 0.5*(Z(IM,K-1)+Z(IM,K))
      XM = 0.5*(X(IM-1,K-1)+X(IM-1,K))
      ZM = 0.5*(Z(IM-1,K-1)+Z(IM-1,K))
      SLG = -(ZM-ZS)/(XM-XS)
      VTOTG = SQRT(VX(IM,K)**2+VZ(IM,K)**2)
      IF (SL.EQ.SLG) GO TO 40
      XG = 0.25*(X(IM,K-1)+X(IM,K)+X(IM-1,K-1)+X(IM-1,K))
      ZG = 0.25*(Z(IM,K-1)+Z(IM,K)+Z(IM-1,K-1)+Z(IM-1,K))
      XB = 2*XS-XG
      ZB = 2*ZS-ZG
      IF (SL.LT.SLG) KN = K-1
      IF (SL.GT.SLG) KN = K+1
      XN = 0.25*(X(IM,KN-1)+X(IM,KN)+X(IM-1,KN-1)+X(IM-1,KN))
      ZN = 0.25*(Z(IM,KN-1)+Z(IM,KN)+Z(IM-1,KN-1)+Z(IM-1,KN))
      VTOTN = SQRT(VX(IM,KN)**2+VZ(IM,KN)**2)
      RATIO = ((ZB-ZG)+SL*(XB-XG))/((ZN-ZG)+SL*(XN-XG))
      P(IM+1,K) = P(IM,K) + RATIO*(P(IM,KN)-P(IM,K))
      RHO(IM+1,K) = RHO(IM,K) + RATIO*(RHO(IM,KN)-RHO(IM,K))
      VTOT = VTOTG + RATIO*(VTOTN-VTOTG)
      E(IM+1,K) = E(IM,K) + RATIO*(E(IM,KN)-E(IM,K))
      GO TO 50
 40   P(IM+1,K) = P(IM,K)
      RHO(IM+1,K) = RHO(IM,K)
      VTOT = VTOTG
      E(IM+1,K) = E(1,K)
 50   VX(IM+1,K) = VTOT*SL/SQL
 60   VZ(IM+1,K) = VTOT/SQL
      RETURN
      END
```
****

135

```
*********************************************************************
*                                                                   *
         SUBROUTINE CURWALL
*                                                                   *
*********************************************************************
C
c.....This subroutine applies the wall boundary condition for a curved surface.
c.....This will be characterized by a normal pressure gradient given by Equation (3.133).
C
         PARAMETER (M=32,M1=33,N=64,N1=65)
         COMMON /BLK0/ IM,KM,KLE,KTE
         COMMON /BLK1/ X(0:M,0:N),Z(0:M,0:N)
         COMMON /BLK4/ RHO(0:M1,0:N1),RHOT(0:M1,0:N1),DRHO(M,N)
         COMMON /BLK8/ P(0:M1,0:N1),RH(0:M1,0:N1)
         COMMON /BLK9/ VX(0:M1,0:N1),VZ(0:M1,0:N1)
         COMMON /BLK10/ E(0:M1,0:N1),H(0:M1,0:N1),T(0:M1,0:N1)
         COMMON /BOUND2/ RCS(N),RCP(N)
C
C-------------------------------------------------------------------
C.....    THE SUCTION SURFACE
         DO 20 K = KLE,KTE
         SL = (X(0,K)-X(0,K-1))/(Z(0,K)-Z(0,K-1))
         SQL = SQRT(1.+SL**2)
         XS = 0.5*(X(0,K-1)+X(0,K))
         ZS = 0.5*(Z(0,K-1)+Z(0,K))
         XM = 0.5*(X(1,K-1)+X(1,K))
         ZM = 0.5*(Z(1,K-1)+Z(1,K))
         SLG = -(ZM-ZS)/(XM-XS)
         RHO(0,K) = RHO(1,K)
         P(0,K) = P(1,K)
         E(0,K) = E(1,K)
         VTOTG = SQRT(VX(1,K)**2+VZ(1,K)**2)
         VTOT = VTOTG
         XX = XG
         ZZ = ZG
         IF (SL.EQ.SLG) GO TO 10
         XG = 0.25*(X(0,K-1)+X(0,K)+X(1,K-1)+X(1,K))
         ZG = 0.25*(Z(0,K-1)+Z(0,K)+Z(1,K-1)+Z(1,K))
         VTOTG = SQRT(VX(1,K)**2+VZ(1,K)**2)
         XB = 2*XS-XG
         ZB = 2*ZS-ZG
         IF (SL.GT.SLG) KN = K-1
         IF (SL.LT.SLG) KN = K+1
         XN = 0.25*(X(0,KN-1)+X(0,KN)+X(1,KN-1)+X(1,KN))
         ZN = 0.25*(Z(0,KN-1)+Z(0,KN)+Z(1,KN-1)+Z(1,KN))
         VTOTN = SQRT(VX(1,KN)**2+VZ(1,KN)**2)
         RATIO = ((ZB-ZG)+SL*(XB-XG))/((ZN-ZG)+SL*(XN-XG))
```

136

```
      RHO(0,K) = RHO(0,K) + RATIO*(RHO(1,KN)-RHO(1,K))
      P(0,K) = P(0,K) + RATIO*(P(1,KN)-P(1,K))
      E(0,K) = E(0,K) + RATIO*(E(1,KN)-E(1,K))
      VTOT = VTOT + RATIO*(VTOTN-VTOTG)
      XX = XX + RATIO*(XN-XG)
      ZZ = ZZ + RATIO*(ZN-ZG)
   10 DN = SQRT((XX-XB)**2+(ZZ-ZB)**2)
      DP = RHO(0,K)*VTOT**2
      IF (SL.GT.SLG) DP = DP*RCS(K-1)
      IF (SL.LT.SLG) DP = DP*RCS(K)
      IF (SL.EQ.SLG) DP = DP*(RCS(K-1)*RCS(K))/2
      P(0,K) = P(0,K) - 2*DP*DN
      VX(0,K) = VTOT*SL/SQL
   20 VZ(0,K) = VTOT/SQL
C
C------------------------------------------------------------------------------------------
C.....  THE PRESSURE SURFACE
      DO 40 K = KLE,KTE
      SL = (X(IM,K)-X(IM,K-1))/(Z(IM,K)-Z(IM,K-1))
      SQL = SQRT(1.+SL**2)
      XS = 0.5*(X(IM,K-1)+X(IM,K))
      ZS = 0.5*(Z(IM,K-1)+Z(IM,K))
      XM = 0.5*(X(IM-1,K-1)+X(IM-1,K))
      ZM = 0.5*(Z(IM-1,K-1)+Z(IM-1,K))
      SLG = -(ZM-ZS)/(XM-XS)
      RHO(IM+1,K) = RHO(IM,K)
      P(IM+1,K) = P(IM,K)
      E(IM+1,K) = E(IM,K)
      VTOTG = SQRT(VX(IM,K)**2+VZ(IM,K)**2)
      VTOT = VTOTG
      XX = XG
      ZZ = ZG
      IF (SL.EQ.SLG) GO TO 30
      XG = 0.25*(X(IM,K-1)+X(IM,K)+X(IM-1,K-1)+X(IM-1,K))
      ZG = 0.25*(Z(IM,K-1)+Z(IM,K)+Z(IM-1,K-1)+Z(IM-1,K))
      XB = 2*XS-XG
      ZB = 2*ZS-ZG
      IF (SL.LT.SLG) KN = K-1
      IF (SL.GT.SLG) KN = K+1
      XN = 0.25*(X(IM,KN-1)+X(IM,KN)+X(IM-1,KN-1)+X(IM-1,KN))
      ZN = 0.25*(Z(IM,KN-1)+Z(IM,KN)+Z(IM-1,KN-1)+Z(IM-1,KN))
      VTOTN = SQRT(VX(IM,KN)**2+VZ(IM,KN)**2)
      RATIO = ((ZB-ZG)+SL*(XB-XG))/((ZN-ZG)+SL*(XN-XG))
      RHO(IM+1,K) = RHO(IM+1,K) + RATIO*(RHO(IM,KN)-RHO(IM,K))
      P(IM+1,K) = P(IM+1,K) + RATIO*(P(IM,KN)-P(IM,K))
      E(IM+1,K) = E(IM+1,K) + RATIO*(E(IM,KN)-E(IM,K))
      VTOT = VTOT + RATIO*(VTOTN-VTOTG)
```

```
            XX = XX + RATIO*(XN-XG)
            ZZ = ZZ + RATIO*(ZN-ZG)
      30    DN = SQRT((XX-XB)**2+(ZZ-ZB)**2)
            DP = RHO(IM+1,K)*VTOT**2
            IF (SL.LT.SLG) DP = DP*RCP(K-1)
            IF (SL.GT.SLG) DP = DP*RCP(K)
            IF (SL.EQ.SLG) DP = DP*(RCP(K-1)+RCP(K))/2
            P(IM+1,K) = P(IM+1,K) - 2*DP*DN
            VX(IM+1,K) = VTOT*SL/SQL
      40    VZ(IM+1,K) = VTOT/SQL
            RETURN
            END
C
*****************************************************************
*
                                                                 *
            SUBROUTINE OUTLET
*
                                                                 *
*****************************************************************
C
c.....   This subroutine applies the outlet boundary condition in accordance with the
c.....   explanation given on page 92.
C
            PARAMETER (M=32,M1=33,N=64,N1=65)
            COMMON /BLK0/ IM,KM,KLE,KTE
            COMMON /BLK4/ RHO(0:M1,0:N1),RHOT(0:M1,0:N1),DRHO(M,N)
            COMMON /BLK8/ P(0:M1,0:N1),RH(0:M1,0:N1)
            COMMON /BLK9/ VX(0:M1,0:N1),VZ(0:M1,0:N1)
            COMMON /BLK10/ E(0:M1,0:N1),H(0:M1,0:N1),T(0:M1,0:N1)
            COMMON /CONST/ CP,CV,GAMMA,GASCON
            DO 20 I = 1,IM
            RHO(I,KM+1) = 2*RHO(I,KM)-RHO(I,KM-1)
            VX(I,KM+1) = 2*VX(I,KM)-VX(I,KM-1)
            VZ(I,KM+1) = 2*VZ(I,KM)-VZ(I,KM-1)
            IF (ISUPER.NE.1) GO TO 10
            P(I,KM+1) = 2.*P(I,KM)-P(I,KM-1)
      10    CONTINUE
            T(I,KM+1) = P(I,KM+1)/(GASCON*RHO(I,KM+1))
            HTV = 0.5*(VX(I,KM+1)**2+VZ(I,KM+1)**2)
            H(I,KM+1) = CP*T(I,KM+1) + HTV
            E(I,KM+1) = CV*T(I,KM+1) + HTV
      20    CONTINUE
            RETURN
            END
C
*****************************************************************
```

References

Numbers in square brackets refer to pages in the present work where these references are cited.

Anderson (Dale A), **Tannehill** (John C), and **Pletcher** (Richard H)
  1984:   *Computational Fluid Mechanics and Heat Transfer*
            Hemisphere Publishing Corporation, New York
  Comments: This book amounts to a comprehensive course in CFD starting from model
            problems such as the wave and heat equations and moving step by step up the
            ladder all the way to the Navier-Stokes equations. The authors are professors
            at Iowa State University.
[see pages 8, 9 (footnote 8), 11, 44]

Burgers (J.M)
  1948:   *A Mathematical Model Illustrating the Theory of Turbulence*
            Advances in Applied Mechanics, Vol 1, pp 171-199
  Comments: A pioneering study of a model equation combining nonlinearity with viscosity
            or heat-diffusion.
[see page 9]

Cajori (Florian)
  1919:   *A History of Mathematics*
            Macmillan Company, New York
  Comments: A useful reference book for the historically-minded student.
[see page 102 (footnote 71)]

Cambel (Ali Bulent) and **Jennings** (Burgess H)
  1958:   *Gas Dynamics*
            McGraw Hill Series in Mechanical Engineering
            McGraw Hill Book Co, New York
  Comments: Although not quite so well-known as the Liepmann/Roshko work on the
            subject, this book includes more practical examples. However no new edition
            has emerged in the computer era, and it now makes somewhat old-fashioned
            reading. The authors were heads of gasdynamics and air-conditioning institutes
            at Illinois and Ohio.
[see page 56]

Conte (S.D) and de Boor (Carl)

1965:   *Elementary Numerical Analysis: an algorithmic approach*
          McGraw Hill Book Company, New York

Comments: An easy-to-understand text-book with plenty of Fortran programs supplementing
          the numerical algorithms. At the time it was written, Fortran had only just been
          developed and the book was very popular with students learning programming.
          The authors were faculty members at Purdue University, Indiana.

[see page 18]

Courant (Richard) and Friedrichs (Kurt Otto)

1948:   *Supersonic Flow and Shock Waves*
          Interscience Publishers Inc, New York

Comments: This was the first of several classic books to emerge from the Institute of
          Mathematics and Mechanics, New York (later renamed the Courant Institute):
          books which place physical problems in a rigorous mathematical setting. The
          current stress on computational methods has somewhat diminished its
          importance, but considered as a purely theoretical work in its area, it is
          probably still unsurpassed.

[see page 62 (footnote 50)]

Courant (R), Friedrichs (K.O) and Lewy (H)

1928:   *Uber die Partiellen Differenzgleichungen*
          der Mathematischen Physik
          Mathematische Annalen, Vol 100, pp 32-74
1967:   *On the Partial Differential Equations of Mathematical Physics*
          (English translation of the 1928 paper)
          IBM Journal, pp 215-34

Comments: This is the paper generally regarded as having laid the cornerstone to the
          theoretical studies of consistency, convergence and stability, which are still very
          much a part of current research.

[see page 62 (footnote 51)]

Courant (Richard) and Hilbert (David)

1924:   *Methoden der Mathematischen Physik*
          Verlag von Julius Springer
1953:   *Methods of Mathematical Physics* .
          Interscience Publishers Inc, New York

Comments: For about half a century, from its first appearance in 1924, this has been the
          definitive work on mathematical methods arising from physical problems. Both
          authors have entries in the Encyclopaedia Britannica.

[see page 59 (footnote 45)]

Denton (J.D)

1974:   *A Time-Marching Method for Two and Three-Dimensional*
Blade to Blade Flows
A.R.C. - R.M. - 3775

Comments: Considered the first effective application of time-marching to a turbomachinery problem. The author is currently a professor at Whittle Laboratories, Cambridge.

[see page 1]

Duffy (Dean G)

1986:   *Solution of Partial Differential Equations*
TAB Books Inc, Pennsylvania

Comments: A book oriented towards practical problems - especially those occurring in nature. The author lectured at the US Naval Academy.

[see page 11]

Farlow (Stanley J)

1982:   *Partial Differential Equations for Scientists and Engineers*
John Wiley and Sons Inc

Comments: An easy-to-read book, each chapter serving as a 'lesson' with a purpose. The author is a professor at the University of Maine.

[see page 11]

Garabedian (Paul R)

1964:   *Partial Differential Equations*
John Wiley and Sons Inc

Comments: A classic text book virtually covering the state-of-the-art knowledge of PDEs prior to the development of CFD. The author was a professor at the Courant Institute of Mathematical Sciences, New York.

[see pages 8, 11 (footnote 11), 13 (footnote 14), 66 (footnote 52)]

Godunov (S.K)

1959:   *A Finite-Difference Method for Numerical Computation of*
Discontinuous Solutions of the Equations of Fluid Dynamics
Mat Sbornik, Vol 47, No 3, pp 271-306

Comments: One of the few Russian papers that are very well-known in the West.

[see page 9]

Guderley (K.G)

1962:   *The Theory of Transonic Flow*
Pergamon Press Inc.

Comments: An overview of theoretical aspects of transonics covering much that was known prior to the CFD era.

[see page 13 (footnote 14)]

Gustafson (Karl E)

1980:   *Introduction to Partial Differential Equations*
          and Hilbert Space Methods
          John Wiley and Sons Inc

Comments: An interesting book especially for the historically-inclined student. Its approach is original and it abounds in footnotes. The author is on the faculty at the University of Colorado.

[see page 11]

Habashi (W.G)

1985:   *Advances in Computational Transonics*
          Pineridge Press, Swansea, Wales

Comments: This is a collection of articles on computational transonics generally involving potential or Euler solvers. The editor is on the faculty at Concordia University, Montreal.

[see page 14 (footnote 15)]

Hirsch (Charles)

1990:   *Numerical Computation of Internal and External Flows*
          John Wiley and Sons Inc

Comments: A CFD reference book in two volumes which starts from the basics and ends up by summarising modern research techniques for the Euler and Navier-Stokes equations. The author is a professor at Vrije University, Brussels and is internationally renowned in his field.

[see pages 14 (footnote 15), 60, 61 (footnote 48), 82, 90]

Jacobi (Karl Gustav)

1841:   *De Formatione et Proprietatibus Determinantium*
          (Concerning the Structure and Properties of Determinants)
          Journal fur die Reine und Angewandte Mathematik
          (Crelle's Journal)

Comments: One of several pioneering studies by this great German mathematician best-known for his fundamental research in elliptic functions.

[see page 43 (footnote 37)]

Jameson (Antony)

1982:   *Transonic Aerofoil Calculations using the Euler Equations*
          Numerical Methods in Aeronautical Fluid Dynamics,
          (Ed: P.L. Roe), Academic Press, London, pp 289-308
          (Proceedings of a Conference held at Reading University in 1981)

Comments: An informal presentation of the author's work. The author - currently a professor at Princeton University - has strong claims to be regarded as the world's leading computational fluid dynamicist.

[see page 77]

142

Jameson (Antony) and **Baker** (Timothy J)

1983:   *Solution of the Euler Equations for Complex Configurations*
        AIAA Paper 81-1259 (6th CFD Conference)
1984:   *Multi-grid solution of the Euler Equation*
        for Aircraft Configurations
        AIAA Paper 84-0093 (22nd Aerospace Sciences Meeting)
[see page 90 (footnote 64)]

Jameson (Antony), **Schmidt** (Wolfgang) and **Turkel** (Eli)

1981:   *Numerical Solution of the Euler Equations by Finite Volume*
        Methods using Runge-Kutta Time-Stepping Schemes
        AIAA Paper 81-1259 (5th CFD Conference)
Comments: This is the paper which originally presented many of the ideas that are made
          use of in the current work. It is probably the first paper to use the Runge-Kutta
          method in the context of a CFD problem.

Kreyszig (Erwin)

1988:   *Advanced Engineering Mathematics*
        John Wiley & Sons, New York
Comments: A fairly comprehensive work covering the elements of most branches of applied
          mathematics including differential equations, linear algebra, vector calculus,
          complex analysis, statistics, and numerical analysis. It has been through six
          editions. The author is a professor of mathematics at Ohio State University,
          Columbus, Ohio.
[see page 54 (footnote 41)]

Lambert (J.D)

1973:   *Computational Methods in Ordinary Differential Equations*
        John Wiley & Sons, New York
Comments: A fairly elementary book: useful for readers who want a numerical background
          in ODEs before moving on to PDEs and CFD. The author was a Reader at the
          University of Dundee, Scotland.
[see page 90]

Lapidus (Leon) and **Pinder** (George F)

1982:   *Numerical Solution of Partial Differential Equations*
        in Science and Engineering
        John Wiley and Sons Inc
Comments: A fairly broad survey of numerical solutions to PDEs using both finite-
          difference and finite-element solvers. The authors are on the faculty at
          Princeton University.
[see page 8]

Lax (Peter D)

1954:   *Weak solutions of nonlinear hyperbolic equations
and their numerical computation
Communications in Pure and Applied Mathematics,
Vol 7, pp 159-193*

1973:   *Hyperbolic Systems of Conservation Laws and the
Mathematical Theory of Shock Waves
Society for Industrial and Applied Mathematics, Philadelphia*

1974:   *Hyperbolic Systems of Conservation Laws II
Communications in Pure and Applied Maths, Vol 10, pp 537-566*

Comments: Very highly rated works from the theoretical point of view though perhaps not essential for the engineering student. The author, one of the world's leading applied mathematicians is currently dean of the Courant Institute of Mathematical Sciences, New York.

[see page 11 (footnote 11)]

Lax (P.D) and **Wendroff** (B)

1960:   *Systems of Conservation Laws
Communications in Pure and Applied Mathematics,
Vol 13, pp 217-37*

Comments: Hirsch (1990) has this to say: 'This remarkable paper contains many basic ideas and considerations which are still highly up to date and we strongly recommend a careful reading of this work'.

[see page 76]

LeVeque (Randall J)

1990:   *Numerical Methods for Conservation Laws
Birkhauser Verlag, Basel*

Comments: A recent monograph on a subject highly relevant to time-marching. The author is on the faculty at the University of Washington, Seattle.

[see page 11]

Liepmann (H.W) and **Roshko** (A)

1957:   *Elements of Gasdynamics
Galicit Aeronautical Series, John Wiley & Sons Inc, New York*

Comments: This was probably the best-known American text book on gasdynamics in the 1950s and 1960s, its authors being on the faculty of the prestigious California Institute of Technology. However with the shift in emphasis to computational (rather than theoretical) methods during the last two decades, it is now a little outdated.

[see page 56]

Lobo (Michael)
1985:   *Stability Theorems Concerning High Order Explicit Algorithms*
          for the Linear Advection Equation
          Cranfield College of Aeronautics Report No 8517
Comments: A summary of research carried out by the present author during the one year
          he spent as a Commonwealth Scholar at the Cranfield College of Aeronautics.
          It turned out that the main theorem had already been proved by Strang (1962),
          though the method used here was quite different and made use of formulae
          from combinatorics.
[see page 60]

Lobo (Michael)
1992:   *Techniques for Accelerating Iterative Schemes*
          Arising from Problems in Flow Modelling
          Cranfield Institute of Technology Research Monograph
Comments: An exposition of the author's research in this area. Though acceleration is the
          central theme of the work, it covers a fairly wide range of problems and
          presents a selection of computer programs that a student may find useful.
[see page 44, 111]

MacCormack (Robert W)
1993:   *A Perspective on a Quarter Century of CFD Research*
          11th AIAA Computational Fluid Dynamics Conference,
          Orlando, Florida, pp 1-15
Comments: This is the keynote lecture at the 1993 AIAA Conference on CFD. The author
          (a pioneer of finite volume methods: see below) is currently a professor of
          aeronautics and astronautics at Stanford University. The article reads like a
          mini-autobiography, describing the author's personal involvement in the
          development of CFD.
[see page 68 (footnote 53)]

MacCormack (R.W) and **Baldwin** (B.S)
1975:   *A Numerical Method for Solving the Navier-Stokes Equations*
          with Application to Shock-Boundary Layer Interactions
          AIAA Paper 72-154
Comments: The paper generally regarded as the first to use artificial viscosity in the context
          of the Euler / Navier Stokes equations.
[see page 76]

MacCormack (R.W) and **Paullay** (A.J)
1972:   *Computational Efficiency Achieved by Time Splitting*
          of Finite-Difference Operators
          AIAA Paper 72-154
Comments: One of two papers (see the next reference) credited with the introduction of the
          finite-volume approach in CFD.
[see page 68 (footnote 53)]

Magnus (R.) and **Yoshihara** (H.)
   1970:   *Inviscid Transonic Flow over Airfoils*
               AIAA Journal, Vol 8, pp 2157-2162

Comments: A pioneering work on the use of time-marching in a CFD problem. With computational resources then available, the method had to be restricted to inviscid two-dimensional flows with the additional assumption of isentropicity to relate pressure and density, thereby dispensing with the need for the energy equation.

McDonald (P.W)
   1971:   *The Computation of Transonic Flow through through*
               Two-Dimensional Gas Turbine Cascades
               ASME Paper 71-GT-89

Comments: One of two papers (see the preceding reference) credited with the introduction of the finite-volume approach in CFD. The author was on the research team at Pratt & Whitney.
[see page 68 (footnote 53)]

Moretti (Gino) and **Abbett** (Michael)
   1966:   *A Time-Dependent Computational Method for Blunt Body Flows*
               AIAA Journal, Vol 4, pp 2136-2141

Comments: This paper is of historical significance in the sense of being the first effective application of time-marching to a CFD problem, although the idea had been suggested as far back as 1950 in the classic paper by Von Neumann and Richtmyer. The Moretti-Abbett paper was also of importance in its time, as it provided a practical means of solving the hypersonic blunt-body problem (a major research problem of the 1960s).
[see page 1]

Murman (E.M) and **Cole** (J.D)
   1971:   *Calculation of Plane Steady Transonic Flows*
               AIAA Journal, Vol 91, pp 114-121

Comments: The paper that made a breakthrough in transonic computations. At the time the authors were working at Boeing Scientific Research Laboratory, Seattle, Washington. Murman has since moved to M.I.T and is the head of the department of aeronautics and astronautics. Cole has moved to Rennslaer Polytechnic Institute, Troy, New York, where he co-authored a book *Transonic Aerodynamics* in 1986
[see page 76]

146

Oswatitsch (Klaus)
   1956:   *Gas Dynamics*
            (English version by Gustav Kierti)
            Academic Press Inc, New York
   Comments: In its day, this was a standard reference book, its author - the director of the
            Institute of Theoretical Gas Dynamics, Aachen - being an authority on the
            subject. Today, however, a standard reference book would be something more
            along the lines of Hirsch.
   [see page 56]

Patankar (Suhas V)
   1980:   *Numerical Heat Transfer and Fluid Flow*
            McGraw Hill, New York
   Comments: A very popular text book for the CFD student learning the ideas behind the
            pressure-correction technique. The author, who wrote this work while at
            Imperial College, London, is currently a professor of Mechanical Engineering
            at the University of Minnesota.
   [see pages 14, 23 (footnote 20)]

Patankar (S.V) and **Spalding** (D.B)
   1972:   *A Calculation Procedure for Heat, Mass and Momentum Transfer
            in Three-Dimensional Parabolic Flows*
            Int. Journal of Heat and Mass Transfer, Vol 15, pp 1787-1806
   Comments: The paper which introduced the pressure-correction technique. Professor
            Spalding is currently the managing director of Concentration Heat and
            Momentum Ltd, Wimbledon.
   [see pages 14, 104 (footnote 73)]

Phillips (G.M) and **Taylor** (P.J)
   1973:   *Theory and Application of Numerical Analysis*
            Academic Press, London & New York
   Comments: An introductory text book - the emphasis being on combining practical methods
            with mathematical theory. It may seem a little old-fashioned to the modern
            student as the authors intended it for a wide range of students including those
            who had no easy access to a computer. In 1973 access to a computer was still
            something of a luxury! The authors were faculty members at the Scottish
            universities of Saint Andrews and Stirling.
   [see page 18]

Ralston (Anthony) and **Rabinowitz** (Philip)
   1978:   *A First Course in Numerical Analysis* (2nd edition)
            McGraw-Hill Book Company, New York
   Comments: A fairly comprehensive tour through basic numerical methods.
   [see page 79]

Richardson (L.F)

1910:   *The Approximate Arithmetical Solution by Finite Differences of*
Physical Problems Involving Differential Equations with an
Application to Stresses in a Masonry Dam
Philosophical Transactions of the Royal Society of London,
Series A, Vol 210, pp 307-357

Comments: A pioneering work regarded as heralding the dawn of CFD The author was a World War I ambulance driver (in the battlefield) and worked on numerical analysis during his spare time.

[see page 1 (footnote 1)]

Richtmyer (Robert D) and Morton (K.W)

1967:   *Difference Methods for Initial-Value Problems*
Interscience Publishers, New York

Comments: An important work which puts numerical solution techniques on a sound mathematical footing. A blend of the abstract and the practical. The authors are leading mathematicians on opposite sides of the Atlantic (Richtmyer at Colorado, Morton at Oxford).

[see pages 4, 11, 64, 66 (footnote 52)]

Rizzi (A.W) and Inouye (M)

1973:   *Time-Split Finite-Volume Method for Three-Dimensional*
Blunt-Body Flow
AIAA Journal, Vol 11, pp 1478-85

Comments: Apparently the first paper use the finite-volume approach in a 3-D setting. The authors were then at NASA Ames, California, though Rizzi subsequently joined the Aeronautical Research Institute, Sweden.

[see page 68 (footnote 53)]

Roberts (G.O)

1971:   *Computational Meshes for Boundary-Layer Problems*
Proceedings of the Second International Conference
on Numerical Methods in Fluid Dynamics
Lecture Notes in Physics, Vol 8, Springer Verlag, New York

Comments: A well-referred-to paper in the context of stretching transformations.

[see page 34]

Roe (Philip L)

1981:   *Numerical Algorithms for the Linear Wave Equation*
          R.A.E. Technical Report 81047

1982:   *Numerical Modelling of Shockwaves and other Discontinuities*
          In *Numerical Methods in Aeronautical Fluid Dynamic,*
          (Ed: P.L. Roe), Academic Press, London, pp 271-93
          (Proceedings of a Conference held in Reading University in 1981)

Comments: These are just two of several papers by the author - an internationally renowned
          computational fluid dynamicist who worked at RAE Bedford and the Cranfield
          Institute of Technology before taking up his present appointment as professor
          at the University of Michigan.

[see pages 8, 9]

Smith (G.D)

1965:   *Numerical Solution of Partial Differential Equations:*
          Finite Difference Methods
          Oxford University Press (2nd edition, 1978: 3rd edition, 1985)

Comments: An easy-to-read book which covers most basic aspects of the subject and
          prepares the student for more advanced works. The author is on the faculty at
          Brunel University.

[see pages 4, 18, 82]

Strang (W. Gilbert)

1962:   *Trigonometric Polynomials and Difference Methods*
          of Maximum Accuracy
          Journal of Mathematics and Physics, Vol 41, pp 147-154

Comments: This paper makes some important contributions to stability theory; in particular,
          it establishes stability criteria for general explicit algorithms for the linear
          advection equation. As the title of the paper suggests, the proof of the main
          theorem makes use of the theory of trigonometric polynomials. See also Lobo
          (1985).

[see page 60]

Tricomi (F)

1923:   *Sulle equazione lineari alle derivate parziale di 2" ordine,*
          di tipo misto
          Atti Accad. Naz. Lincei, Rend, Vol 14, pp 133-247

Comments: A pioneering treatise on mixed partial differential equations - thereby
          contributing to a theoretical understanding of transonic flow.

[see page 13 (footnote 14)]

Von Neumann (John) and Richtmyer (Robert D)

1950: *A Method for the Numerical Computation of Hydrodynamic Shocks*
Journal of Applied Physics, Vol 21, pp 232-237

Comments: This paper is of historical significance for two reasons. It was the first to suggest the possibility of employing time-dependent techniques to solve asymptotically for steady flows (though computational resources of the time precluded any practical applications of this idea for the next 16 years). It was also the first to propose the all-important concept of artificial viscosity. The first author was one of the world's leading mathematicians and physicists in the first half of the twentieth century and has a sizeable entry in the Encyclopaedia Britannica. The second author, a professor at Colarado, later collaborated with Professor Morton of Oxford on an important text book on numerical methods.

[see pages 10, 74]

Whitham (Gerald Beresford)

1974: *Linear and Nonlinear Waves*
John Wiley and Sons Inc

Comments: Despite the two decades that have elapsed since its publication, this is still probably the most comprehensive single-volume work in its area. The author (a Fellow of the Royal Society) is a professor at the California Institute of Technology.

[see pages 10 (footnote 9), 11]

Wu (C.H)

1952: *A General Theory of Three-Dimensional Flow in Subsonic and Supersonic Turbomachines of Axial, Radial and Mixed Flow Types.*
NACA-TN-2604

Comments: A landmark paper in computational methods in turbomachinery: the paper that introduced the streamsurface technique.

[see page 105]

Zachmanoglou (E.C) and Thoe (Dale W)

1976: *Introduction to Partial Differential Equations with Applications*
Dover Publications Inc, New York

Comments: One of the few easy-to-read works on PDEs which stress more towards the theoretical aspects. The authors are faculty members at the University of Purdue.

[see pages 8, 11, 12]